A SURVEY OF ATTITUDES AND ACTIONS ON DUAL USE RESEARCH IN THE LIFE SCIENCES

A Collaborative Effort of the National Research Council and the American Association for the Advancement of Science

Committee on Assessing Fundamental Attitudes of Life Scientists
as a Basis for Biosecurity Education

Development, Security and Cooperation
Policy and Global Affairs

Board on Life Sciences
Division on Earth and Life Studies

NATIONAL RESEARCH COUNCIL
OF THE NATIONAL ACADEMIES

In collaboration with

AAAS
ADVANCING SCIENCE, SERVING SOCIETY

THE NATIONAL ACADEMIES PRESS
Washington, D.C.
www.nap.edu

THE NATIONAL ACADEMIES PRESS 500 Fifth Street, N.W. Washington, DC 20001

NOTICE: The project that is the subject of this report was approved by the Governing Board of the National Research Council, whose members are drawn from the councils of the National Academy of Sciences, the National Academy of Engineering, and the Institute of Medicine. The members of the committee responsible for the report were chosen for their special competences and with regard for appropriate balance.

This project was supported by the Carnegie Corporation of New York, the Alfred P. Sloan Foundation, and funds provided by the Presidents' Circle Communications Initiative. Any opinions, findings, conclusions, or recommendations expressed in this publication are those of the author(s) and do not necessarily reflect the views of the organizations or agencies that provided support for the project.

International Standard Book Number-13: 978-0-309-12510-9
International Standard Book Number-10: 0-309-12510-3
Library of Congress Control Number: 2009925225

Additional copies of this report are available from the National Academies Press, 500 Fifth Street, N.W., Lockbox 285, Washington, D.C. 20055; (800) 624-6242 or (202) 334-3313 (in the Washington metropolitan area); Internet, http://www.nap.edu

Suggested citation: National Research Council. 2009. A Survey of Attitudes and Actions on Dual Use Research in the Life Sciences: A Collaborative Effort of the National Research Council and the American Association for the Advancement of Science. Washington, D.C.: The National Academies Press.

Printed in the United States of America

THE NATIONAL ACADEMIES

Advisers to the Nation on Science, Engineering, and Medicine

The **National Academy of Sciences** is a private, nonprofit, self-perpetuating society of distinguished scholars engaged in scientific and engineering research, dedicated to the furtherance of science and technology and to their use for the general welfare. Upon the authority of the charter granted to it by the Congress in 1863, the Academy has a mandate that requires it to advise the federal government on scientific and technical matters. Dr. Ralph J. Cicerone is president of the National Academy of Sciences.

The **National Academy of Engineering** was established in 1964, under the charter of the National Academy of Sciences, as a parallel organization of outstanding engineers. It is autonomous in its administration and in the selection of its members, sharing with the National Academy of Sciences the responsibility for advising the federal government. The National Academy of Engineering also sponsors engineering programs aimed at meeting national needs, encourages education and research, and recognizes the superior achievements of engineers. Dr. Charles M. Vest is president of the National Academy of Engineering.

The **Institute of Medicine** was established in 1970 by the National Academy of Sciences to secure the services of eminent members of appropriate professions in the examination of policy matters pertaining to the health of the public. The Institute acts under the responsibility given to the National Academy of Sciences by its congressional charter to be an adviser to the federal government and, upon its own initiative, to identify issues of medical care, research, and education. Dr. Harvey V. Fineberg is president of the Institute of Medicine.

The **National Research Council** was organized by the National Academy of Sciences in 1916 to associate the broad community of science and technology with the Academy's purposes of furthering knowledge and advising the federal government. Functioning in accordance with general policies determined by the Academy, the Council has become the principal operating agency of both the National Academy of Sciences and the National Academy of Engineering in providing services to the government, the public, and the scientific and engineering communities. The Council is administered jointly by both Academies and the Institute of Medicine. Dr. Ralph J. Cicerone and Dr. Charles M. Vest are chair and vice chair, respectively, of the National Research Council.

www.national-academies.org

AAAS

ADVANCING SCIENCE, SERVING SOCIETY

Founded in 1848, AAAS serves some 262 affiliated societies and academies of science, serving 10 million individuals. *Science* has the largest paid circulation of any peer-reviewed general science journal in the world, with an estimated total readership of one million. The nonprofit AAAS is open to all and fulfills its mission to "advance science and serve society" through initiatives in science policy, international programs, science education, and more. EurekAlert!, the premier science-news Web site, a service of AAAS, provides the latest research news.

COMMITTEE ON ASSESSING FUNDAMENTAL ATTITUDES OF LIFE SCIENTISTS AS A BASIS FOR BIOSECURITY EDUCATION

Ronald M. Atlas, Ph.D., (*Chair*), Professor of Biology and Public Health, and Co-director of the Center for Health Hazards Preparedness, University of Louisville, Louisville, KY
Robert Cook-Deegan, M.D., Director of the Institute for Genome Sciences, and Policy Center for Genome Ethics, Law & Policy, Duke University, Durham, NC
David Franz, D.V.M., Ph.D., Chief Biological Scientist, Midwest Research Institute, Frederick, MD
James Lepkowski, Ph.D., Research Professor, Institute for Social Research; Professor, Department of Biostatistics; University of Michigan, Ann Arbor
Francis Macrina, Ph.D., Edward Myers Professor of Dentistry and Vice President for Research, Virginia Commonwealth University, Richmond, VA
Kathleen Vogel, Ph.D., Assistant Professor, Department of Science and Technology Studies and the Peace Studies Program, Cornell University, Ithaca, NY

Staff

Kavita Marfatia Berger, Ph.D., Project Director, Center for Science, Technology, and Security Policy, American Association for the Advancement of Science, Consultant
Kerry Brenner, Ph.D., Senior Program Officer, Board on Life Sciences
Jo L. Husbands, Ph.D., Senior Project Director, Program on Development, Security and Cooperation
Fran Sharples, Ph.D., Director, Board on Life Sciences
John Sislin, Ph.D., Study Director
Rae Allen, Administrative Assistant

Preface and Acknowledgments

To begin to assess the attitudes of members of the life sciences community and to learn what actions life scientists in the United States would support to reduce the risk of misuse of the results of scientific research, the National Research Council (NRC) of the National Academies and the American Association for the Advancement of Science (AAAS) conducted a survey of life scientist members of AAAS to assess their *awareness* of the dual use dilemma, including their perceived risks of bioterrorism, and their *attitudes* toward their responsibilities to help reduce the risks that their research could be misused.

NRC appointed a committee to provide oversight for the Academies' participation in the project. The committee was not formally appointed until after the stages of the project that developed the survey instrument and survey design and decided whom to survey. The committee did provide oversight for the analysis of the survey results and the preparation of the final report. The committee is fully responsible for the interpretation of the data.

Funding for the project came from several sources. A generous grant from the Alfred P. Sloan Foundation provided the primary source of funding. In addition, the initial development of the project was supported by a planning grant from The Carnegie Corporation of New York, while the Presidents Circle Communications Initiative of the National Academies supported the focus groups. Fielding of the survey was supported by the AAAS Center for Science, Technology and Security Policy and funds from the John D. and Catherine T. MacArthur Foundation.

The results of the survey provide some of the first empirical data about the awareness and attitudes of a sample of U.S. life scientists toward biosecurity and the potential misuse of legitimate scientific research for malicious purposes, that is, their awareness of and attitudes toward the so-called dual use dilemma. Unfortunately, a low response rate and uncertainties about whether the respondents are representative of the broader U.S. life sciences community limit the ability to generalize from the responses that were obtained from the survey. Nevertheless, the findings are valuable in generating hypotheses that can be tested in future efforts. There appears to be support among life scientists for self-regulatory approaches to reducing the risk of misuse of scientific knowledge. In fact the survey results suggest that concerns about dual use research have led some scientists to change their research activities. This may be an indication that the life sciences community is responsibly responding to reduce the risk of misuse of science. But it is also possible that some scientists are overreacting to the perceived threat, for example, by breaking collaborations and excluding foreigners from their laboratories. The committee feels that it is important to investigate further what changes are being made in the conduct of research by U.S. researchers in response to dual use concerns and how this may be impacting the conduct of research in the life sciences.

As recognized in previous NRC reports, notably *Biotechnology Research in an Age of Terrorism*, the committee feels that it is important for all involved to recognize that protection of the life sciences against misuse requires a global effort. This survey was a first step to learn about the level of awareness of dual use research within the life science community in the United States and the policy measures that would be supported by that community. Future efforts will need to assess the prospect of the international life science community accepting various policy proposals aimed at reducing the risks of misuse of legitimate life science research. The committee hopes that its work will help further the essential dialogue and empower the voices of the life sciences community to be heard.

This report has been reviewed in draft form by individuals chosen for their diverse perspectives and technical expertise, in accordance with procedures approved by the National Academies' Report Review Committee. The purpose of this independent review is to provide candid and critical comments that will assist the institution in making its published report as sound as possible and to ensure that the report meets institutional standards for objectivity, evidence, and responsiveness to the study charge. The review comments and draft manuscript remain confidential to protect the integrity of the process.

We wish to thank the following individuals for their review of this report: Gerald Epstein, Center for Strategic and International Studies;

Rachel Levinson, Arizona State University; Filippa Lentzos, London School of Economics; Nancy Mathiowetz, University of Wisconsin; Henry Metzger, National Institute of Arthritis and Musculoskeletal and Skin Diseases; Stephen Morse, Columbia University; Victoria Sutton, Texas Tech University; and Judy Tanur, The State University of New York.

Although the reviewers listed above have provided many constructive comments and suggestions, they were not asked to endorse the conclusions or recommendations, nor did they see the final draft of the report before its release. The review of this report was overseen by Edward Perrin, University of Washington, and Mary Clutter (Retired), National Science Foundation. Appointed by the National Academies, they were responsible for making certain that an independent examination of this report was carried out in accordance with institutional procedures and that all review comments were carefully considered. Responsibility for the final content of this report rests entirely with the authoring committee and the institution.

The committee wants to thank the members of the NRC and AAAS staff who provided extensive input during the project. The design and development of the project was a group effort among the staff from several parts of the NRC and AAAS. John Sislin served as study director, provided the data and statistical analyses that were the core of this study, as well as drafting the methods and findings chapters. Jo Husbands and Kavita Berger also made substantial contributions to the many drafts of the report. Fran Sharples and Kerry Brenner provided insightful comments that helped guide the committee in its work. We also want to extend our thanks to Connie Citro and Michael Cohen from the NRC Committee on National Statistics for providing additional statistical advice. Finally, the committee wishes to acknowledge the work of the staff at AAAS who were involved in fielding the survey that provided the data for the committee to analyze.

The committee is especially grateful to those who participated in the focus groups that helped formulate the survey questionnaire and the many scientists who took the time to complete the survey. Although filling out surveys is tedious and often unrewarding, the committee hopes that the completion of this report will be valuable to the life sciences community and policy makers who are trying to advance the responsible conduct of science and ensure that knowledge in the life sciences is not misused to do harm.

Ronald M. Atlas, *Chair*

Contents

Appendixes

List of Tables, Figures, and Boxes

TABLES

FIGURES

BOXES

Summary

Over the past 50 years, rapidly expanding knowledge in the biological sciences has brought great benefits to society. But the same technologies that fuel scientific advances also pose potential risks—that the knowledge, tools, and techniques gained through legitimate biotechnology research could be misused to create biological weapons or for bioterrorism. This is often called the dual use dilemma of the life sciences. The fear is that some research—dual use research of concern—could be used by those with malicious intent to do great harm. Yet even research with the greatest potential for misuse may offer significant benefits as well. Determining how to constrain the danger without harming essential scientific research is critical for national security as well as prosperity and well-being.

Growing concerns about the potential risks of dual use research of concern have led to calls for raising awareness within the scientific community about dual use issues. Several reports from the National Research Council, including *Biotechnology Research in an Age of Terrorism* (NRC 2004a, herein called the Fink report), *Seeking Security: Pathogens, Open Access, and Genome Databases* (NRC 2004b), and *Globalization, Biosecurity, and the Future of the Life Sciences* (NRC 2006a) share a common message: The scientific community should take preemptive steps to protect the integrity of science and to minimize the risk of misuse of dual use research of concern. These reports also contain recommendations for enhanced education and outreach programs to raise awareness of the potential unintended harm from dual use research. They recommended that scientific societies and professional associations undertake programs to educate

scientists about dual use issues and their responsibilities to help mitigate the risks of misuse.

In addition to proposed efforts by professional and scientific societies, the National Science Advisory Board for Biosecurity (NSABB), created in 2004 in response to the Fink report (NRC 2004a), has an explicit mandate to "provide recommendations on the development of mandatory training programs for education and training in biosecurity issues for all scientists and laboratory workers at federally funded institutions." A few universities, nongovernmental organizations, and professional societies have undertaken or are planning education efforts even before there is any government mandate to do so. For example, though certainly not exhaustive, in the United States the Federation of American Scientists, the Southeast Regional Center of Excellence for Emerging Infections and Biodefense, and the Center for Arms Control and Nonproliferation all have online materials or programs available.

THE AAAS-NRC SURVEY PROJECT

In September 2005, NRC and the American Association for the Advancement of Science (AAAS) Center for Science, Technology, and Security Policy (CSTSP) hosted a meeting, "Education and Raising Awareness: Challenges for Responsible Stewardship of Dual Use Research in the Life Sciences," that brought together over 50 participants to share information and explore ways to engage and educate the research community most effectively. The discussions underscored how little data exist about levels of awareness and attitudes about biosecurity issues in the life sciences community and highlighted the critical need to move beyond anecdotes to empirical evidence.

Building on the results of their 2005 planning meeting, CSTSP and NRC developed a plan to survey a sample of AAAS members in the life sciences about their knowledge of dual use issues and attitudes about their responsibilities to help mitigate the risks of misuse of their research. In addition to providing essential baseline data, it was hoped that the results of the survey would generate more attention to the continuing challenges of dual use issues and foster additional debate among life scientists about their personal and professional responsibilities. The project used consultations with experts and practicing scientists as well as four focus groups in 2007 to design and refine a Web-based survey questionnaire that could be e-mailed to AAAS members in the life sciences.

The focus of the survey was on practicing scientists in the biological, health, and agricultural sciences working in the United States. AAAS is the largest general scientific society in the world and has more than 64,000 life scientists among its members. Since the membership is largely American

(about 84 percent) and primarily composed of scientists with advanced degrees (e.g., Ph.D.s or M.D.s), there was ample opportunity to attempt to survey the attitudes of American researchers across the full range of life science subfields. Because the survey was conducted via e-mail the study was restricted to 24,194 members who had validated e-mail addresses out of 64,787 life scientists who belong to AAAS. A random sample of 10,000 from these 24,194 AAAS members was selected to be contacted.

The survey was fielded from early August to early October 2007 by the AAAS Office of Member Services, with several follow-up e-mails to encourage a higher response rate. Among those sent the survey, 2,713 individuals viewed the survey (i.e., clicked on the link to the questionnaire provided in an e-mail); 1,954 individuals completed part of the survey; and 1,570 completed the entire survey. This leads to a response rate of about 16 percent for completed surveys and 20 percent including partial responses.

Almost all of the respondents had conducted or managed life sciences research (and three-quarters of them are currently doing so), were employed, had a postgraduate degree, and were U.S. citizens. In addition, a substantial majority of the scientists were academics and most were mid-career.

Given the low response rate, the lack of information by which the characteristics of the nonrespondents could be compared to those of the respondents, and the fact that the sampling frame included only those AAAS members whose e-mail addresses were known to AAAS, the survey results should not be generalized to the general population of U.S. life scientists. The methodological difficulties encountered in this project with regard to obtaining a representative sample and a high enough response rate to make generalized conclusions provide valuable lessons for future surveys on this as well as other topics of interest to the scientific community. Although it is necessary, because of these issues, to confine the report to the respondents and not to generalize beyond them, the committee believes that the survey results (including respondents' anecdotal comments) provide interesting indications of how the U.S. life sciences community may view dual use research that merit further investigation.

SURVEY RESULTS

The results of the survey provide some of the first empirical data about the perceptions of a sample of U.S. life scientists across a variety of disciplines about the potential risks of misuse of legitimate scientific research for malicious purposes. The survey data provide evidence about how the respondents perceive the sources of risk related to dual use research, the actions that some of these scientists are taking to reduce the risk of misuse

of science, and the prospects for acceptance of various policy proposals aimed at reducing the risks of misuse of legitimate life science research, although, as indicated earlier, the results of the survey must be viewed with caution because of the low response rate and possible response bias. Scientists who may be involved in biodefense research or who use select agents, for example, may be more aware of the dual use dilemma and thus more likely to have responded to the survey. In addition, a few of the questions could have been interpreted in multiple ways so that, for example, all "Yes" or "No" responses may not be comparable. Despite these potential problems, the committee believes the data obtained in this study offer valuable insights and new information.

Overall, the survey findings suggest that there may be considerable support for models of oversight that rely on the responsible conduct of research and self-governance by the scientific community. The responses also suggest, however, that there is a critical need to clarify the scope of research activities of high concern and to determine the appropriate actions that members of the life sciences community can take to reduce the risk of misuse of science for biological weapons development and bioterrorism.

Perceptions of Risk

The findings suggest that, on average, the scientists who responded to the survey perceive a potential, but not overwhelming, risk of bioterrorism and that the risk is greater outside the United States. On average, the respondents believed that there is a 51 percent chance that there will be an act of bioterrorism somewhere in the world in the next 5 years and a 35 percent chance that there will be an act of bioterrorism in the United States in the next 5 years. Three-quarters of the respondents believe that a preference for other means of attack is the primary reason why there have been only a few acts of bioterrorism to date; overwhelmingly, 87 percent of respondents said that they believe that terrorists are not deterred by the threat of being caught and punished. Fewer scientists considered a lack of knowledge (46 percent) or access to equipment (51 percent) or agents (36 percent) to be significant barriers. It may be that one's perceived risk of such an attack is related to one's support for taking measures to reduce the risks that life sciences research might be misused.

With regard to the chance that the knowledge, tools, or techniques from dual use research will facilitate bioterrorism, the respondents perceive a 28 percent chance, on average, of such a bioterror attack within the next 5 years. Half of the respondents thought that if someone wanted to create a harmful biological agent, the Internet would be the most likely place to provide sufficient information for life scientists with college-

level training. Other sources of information—articles in scientific journals (40 percent), personal communications (38 percent), and presentations at professional meetings (18 percent)—were considered relatively less likely sources, although on average 45 percent of respondents answered "Don't Know" to these questions.

Actions Taken by Life Scientists in Response to Dual Use Concerns

Although the responses to the survey indicate that bioterrorism probably is not perceived to present a serious immediate risk to U.S. or global security, the survey results also indicate that there is already concern about dual use issues among some of the life scientists who responded. Fifteen percent of the respondents (260 individuals out of 1,744) indicated that they are so concerned about dual use research that they have taken actions, even in the absence of guidelines or mandatory regulations from the U.S. government. Some respondents reported that they had broken collaborations, not conducted some research projects, or not communicated research results. The results indicate that more scientists have modified their research activities than some members of the committee expected on the basis of previous reports of manuscripts that have been modified or not published because of dual use concerns.

Interestingly, many of the actions that the respondents reported taking to mitigate concerns occurred before the publication stage; much of the behavior change occurred during the research design, collaboration, and early communication stages. Of particular interest and concern to the committee, a few respondents commented on their concerns about foreigners as potential security risks, which may be reflected in the reported avoidance of some collaborations.

The survey results suggest that: (1) some life scientists in the United States may be willing to consider self-governance aimed at the responsible scientific conduct for dual use research, and (2) some life scientists in the United States are already acting, even in the absence of government regulations and guidance, to protect against the perceived risk of misuse of dual use research.

Oversight Mechanisms

With a proposed oversight framework for dual use research of concern proposed by NSABB in June 2007 now under consideration within the U.S. government, the survey was an opportunity to assess scientists' attitudes toward specific policy options. Many of the respondents indicated that they believe that personal responsibility, including measures such as codes of conduct, could foster a positive culture within the scientific com-

munity to evaluate the potential consequences of their research for public safety and national security. They also indicated that they believe that individual researchers, professional scientific societies, institutions, and scientific journals should be responsible for evaluating dual use potential of research and/or fostering the culture of scientific responsibility.

A majority of those who responded to the survey favored self-governance mechanisms for dealing with dual use research of concern, such as those proposed by the Fink report (NRC 2004a), rather than additional mandatory government regulations. In addition to the low level of support for greater federal oversight (26 percent), the individual comments indicated a belief that increased government oversight of dual use research would be counterproductive by inhibiting the research needed to combat emerging infectious diseases and bioterrorism as well as being potentially harmful to the scientific enterprise more generally.

The survey suggests that most of the respondents (82 percent) favor their professional societies' prescribing a code of responsible conduct to help prevent misuse of life sciences research. However, many respondents (66 percent) did not know whether the societies to which they belonged already had codes that address dual use issues, and some of the societies most frequently cited do not in fact have a code. There was substantially less support (38 percent agree or strongly agree) for a Hippocratic-style oath.

The results also indicate potential support for journals having biosecurity policies. Yet, most of the respondents did not know if any of the journals in which they have published or to which they have submitted manuscripts have those policies. Moreover, more than half of those who responded to the survey strongly disagreed or disagreed with restrictions on personal communication, altering or removing methods or findings from scientific publications, or limiting publication itself.

The survey points to a likely preference for self-governance measures to provide oversight of dual use research. There was substantially less support for mandatory measures that might be imposed by regulation, although the results varied for different policy measures. The results indicate that there may be greater support for restrictions on access to biological agents (just under 50 percent of the respondents said they agree or strongly agree) and certifications of researchers (just over 40 percent of the respondents said they agree or strongly agree) than for any control of scientific knowledge generated from the research or through information exchange (only 20 to 30 percent of respondents supported these measures). Table S-1 provides a list of the level of support for the various measures addressed in the survey.

TABLE S-1 Summary of Results Regarding Support for Measures of Personal and Institutional Responsibility

Measures of Personal or Institutional Responsibility	Strongly Agree or Agree (or Respond Yes*) (%)
Principal investigators should be responsible for the initial evaluation of the dual use potential of their life sciences research.	87
Principal investigators should be responsible for training lab staff, students, and visiting scientists about dual use research.	86
Should professional science societies have codes for the responsible conduct of dual use life sciences research?	82*
University and college students should receive educational lectures and materials on dual use life sciences research.	68
Scientists should provide formal assurance to their institution that they are assessing their work for dual use potential.	67
Funding agencies should require grantees to attest on grant applications that they have considered dual use implications of their proposed research.	60
Should scientific journals have policies regarding publication of dual use research?	57*
Institutions should provide mandatory training for scientists regarding dual use life sciences research.	55
Greater restrictions should be placed on access to specific biological agents or toxins.	47
Researchers conducting dual use research should be certified.	42
All grant proposals for life sciences research with dual use potential should be reviewed by a researcher's institution prior to submission for funding.	41
Scientists conducting or managing research should take an oath.	38
Research findings should be classified based on their dual use potential.	28
Dual use research needs greater federal oversight.	26
Certain experimental methods or findings should be altered or removed prior to publication or presentation.	22
Certain biological equipment that is commonly used in life science research should be licensed.	21
There should be restrictions on disclosure of details about the research or its findings through personal communication.	21
There should be restrictions on publication of findings based on their dual use potential.	21

SOURCE: NRC/AAAS Survey of Life Scientists; data analysis by staff.

The survey results suggest there is *support* for:

1. Greater oversight that is not federally mandated,
2. Self-governance mechanisms as an approach for preventing misuse of life science research and knowledge,
3. Professional and scientific societies adopting codes of conduct that include dual use research as suggested in the Fink report (NRC 2004a),
4. Establishing and implementing policies for authors and reviewers to consider the dual use potential of research manuscripts submitted to journals.

The survey results suggest there is *opposition* to:

1. Mandatory government regulations to govern the conduct of dual use research and the communication of knowledge from that research;
2. Other mandatory oversight actions, such as oaths or licensing of scientists.

Based on the survey results and its own analysis, the committee believes that a basis of support exists within the U.S. scientific community for measures that, taken together, could lead to the development of a system of self-governance for the oversight of key aspects of dual use research.

Education and Outreach

A major reason for conducting the survey was to inform efforts for education and awareness-raising about dual use research by providing empirical data on the attitudes of a sample of the life sciences community. In general, the respondents to this survey would likely support educational and outreach activities aimed at raising awareness of the dual use dilemma. The respondents indicated that they supported educational materials and lectures on dual use research for students. They also supported mandatory training by institutions for practicing life scientists regarding dual use research of concern.

The survey results also highlight the need to better define the scope of dual use research of concern. Fewer than half of the respondents who indicated that they were carrying out dual use research activities felt that their research fell into one of the seven categories of research of concern specified by the NSABB. The dual use experiments of concern as listed in the Fink report (NRC 2004a) and by the NSABB are all based on microbial research, but other relevant research, such as theoretical research, scenario development, or applied research (e.g., pharmaceutical formulations or

neuroscience research) can be of dual use concern. In their individual comments, a number of respondents stressed the difficulties of defining dual use, as did participants in the focus groups used to develop the survey. Clearly a better understanding of the scope of dual use research of real concern would help any educational or outreach activities aimed at raising the awareness of life scientists so that appropriate actions can be taken.

Based on the survey results and its own analysis, the committee believes that there is support for mandatory education and training about dual use issues, most likely as part of ethics and responsible conduct of research training.

RECOMMENDATIONS

The committee believes that the survey raises several hypotheses that merit further research about the views of life scientists about oversight policies and education and outreach efforts to address concerns about dual use issues in the life sciences. In particular, based on the survey results and its own deliberations, the committee offers the following recommendations:

Oversight, Education, and Outreach

1. Explore how to continue and to expand the dialogue within the life sciences community about dual use research of concern.
2. Explore ways to provide guidance to the life sciences community about appropriate actions that can be taken to protect against the misuse of dual use research.
3. Seek to better define the scope of knowledge in the life sciences that may be at greatest risk for misuse and to provide the life sciences community with criteria for recognizing dual use research of concern.
4. Encourage journals that have biosecurity policies or plan to adopt them in the future and the professional and scientific societies that have or plan to develop codes of conduct to communicate those policies more effectively.

Further Research

1. Examine the effectiveness of existing educational programs and how they can be enhanced and focused.
2. Seek to extend educational and awareness-raising efforts being conducted in the United States to the broad international scientific community.

3. Examine how education and outreach activities can be developed to guide the life science community's response to concerns about dual use research so as to ensure that actions taken by the community are appropriate and contribute to advancing scientific knowledge while protecting national security.

4. Conduct additional surveys, interviews, or focus groups of U.S. life scientists that better represent the full community, with higher response rates than the current study was able to achieve, and the ability to assess potential bias, in order to gain

- a better understanding of the awareness of a broader range of U.S. life scientists about dual use research of concern and the measure that they would support to reduce the threat that research in the life sciences could be subverted to do harm;
- a better understanding of the types of behavioral changes being made in response to dual use concerns to determine if actions by life scientists are contributing to national security or harming scientific research; such research is critical given the actions that the current survey suggests are being taken;
- more detailed information about the types of changes scientists are making or scientists' thoughts about dual use issues, experiments of concern, and select agents;
- a better understanding of scientists' experiences with education on this topic and their views about the content and delivery of educational and training materials.

5. Conduct additional surveys of life scientists outside the United States that would enable comparisons of attitudes toward dual use research of concern and inform educational and outreach programs so that they can be effective on a global scale. Such knowledge could also facilitate international discussions of potential measures to address dual use concerns.

Overview

INTRODUCTION

Prior to 2001, the life sciences were largely spared from government concerns about the potential for misuse of scientific knowledge. Only a few in the scientific community had raised concerns about the potential contributions of life sciences research to biological weapons programs and bioterrorism before the anthrax attacks of 2001.[1] There were no regulatory

[1]In his presentation to the May 2000 annual meeting of the National Academy of Sciences, Matthew Meselson, a leading figure in the life sciences and biological arms control communities, highlighted the potential for harm that could be done through biotechnology: "Every major technology—metallurgy, explosives, internal combustion, aviation, electronics, nuclear energy—has been intensively exploited, not only for peaceful purposes but also for hostile ones. Must this also happen with biotechnology, certain to be a dominant technology of the coming century? During the century just begun, as our ability to modify fundamental life processes continues its rapid advance, we will be able not only to devise additional ways to destroy life but will also be able to manipulate it—including the processes of cognition, development, reproduction, and inheritance. A world in which these capabilities are widely employed for hostile purposes would be a world in which the very nature of conflict has radically changed. Therein could lay unprecedented opportunities for violence, coercion, repression, or subjugation" (Meselson 2000). Meselson and others were worried about biological weapons programs, such as those of the former Soviet Union, and what those programs could do with advanced molecular biology to make biological weapons. He and others (e.g., Block 1999; Roberts 2000; Tucker 2000; Leitenberg 2005) were concerned with the intentional use of biotechnology for weapons development and how to evaluate the threat of bioterrorism. However, these studies were not yet focused on the potential subversion of legitimate research in the life sciences for malevolent purposes of bioterrorism. Concerns about bioterrorism risks grew substantially after the first World Trade Center attack, the Oklahoma City

or legislative actions taken by the U.S. government aimed at constraining research and communication in the life sciences outside of some government laboratories where classified biological research was performed.[2] But as fear gripped the nation in the aftermath of the tragic events of September 11 and the anthrax attacks that followed a month later, questions began to be raised about whether publicly available scientific information could be used for malevolent purposes and what actions the government should take to constrain "dangerous" information. The security community and policy makers in the United States began to discuss whether some life science research should be categorized as "sensitive but unclassified (SBU) information," asking whether such information needed to be constrained to protect against future bioterrorist attacks.[3] Additional discussions focused on risks from international collaborations and whether research by foreign graduate students at U.S. academic institutions should be restricted.

Given the high level of anxiety about the anthrax attacks, accentuated by allegations in the news media about who conducted those attacks and uncertainties as to where the anthrax bacteria originated, questions arose

bombing, and the Aum Shinrikyo sarin chemical attack and revelations about the latter's efforts to develop bioweapons (Wright 2007). But much of the discussion focused on the likelihood of terrorist groups' pursuing biological capabilities and their abilities to overcome technical barriers to acquisition and use (for a review of these discussions and debates, see Frerichs et al. [2004]). It was not until after the publication of a paper by Australian researchers showing that the insertion of an interleukin gene, IL-4, into the mousepox virus could render the virus vaccine-resistant (Jackson et al. 2001) that concern about the potential contribution of publications in the open literature to enabling bioterrorism became a significant focus of concern. Advances in biotechnology have been increasingly seen as a dangerous and powerful new way to produce biological weapons.

[2]Regulations were put in place to control the transfer of select microbial pathogens and toxins in 1996, but knowledge generated by fundamental biological research was not viewed with concern.

[3]SBU is one of dozens of categories the federal government uses to control access to information; like many others, it has never been defined in statute. SBU has been used to denote unclassified national security information that might nonetheless be useful to an adversary. Efforts to define what constitutes SBU information provoke recurrent controversies (National Research Council [NRC] 2007a). A review of SBU and other such categories by the Congressional Research Service provides a detailed history (Knezo 2004). On May 7, 2008, the White House announced a new policy to create a single category, "Controlled Unclassified Information," that is to apply across the executive branch (White House 2008). Many of the details of how the new policy will be implemented have not been decided or released.

about the safety of freely disseminating knowledge in the life sciences.[4] Policy makers and members of the security and scientific communities soon began to focus on the dual use dilemma in the life sciences[5]—recognizing that the very research needed for bettering human health, advancing the economy, and other societal benefits could be misused to do harm. Slowing research in the life sciences, however, would harm the nation, global health, and the advancement of science, and so whatever policies might be developed to enhance security needed to be crafted very carefully. Given this tension, it is not surprising that 7 years later the debates continue over what to do about the dual use dilemma for research and communication in the life sciences.[6]

Clearly, during the past 50 years, rapidly expanding knowledge in the life sciences has brought great benefits to society. Smallpox has been eradicated; new vaccines are available to prevent childhood diseases such as measles, mumps, and rubella; there is a vaccine to protect against cervical cancer; numerous therapeutic drugs are available to treat infectious diseases, heart disease, cancer, etc.; and life expectancy has been increasing. Moreover, with regard to national security, research activities in the life sciences are vital for providing essential protection against infectious diseases and bioterrorism through understanding pathogenesis and host–agent (pathogen or toxin) interactions, and the development of vaccines, therapeutics, and diagnostics. In many areas, advances in the life sciences, enabled by government investments such as those by the National Institutes of Health (NIH) have led to economic development in the United States, which contributes to national security and national prosperity.

As a result of its preeminent research enterprise, the United States has achieved a global leadership position in biotechnology. The continu-

---

[4]In August 2008, after the current survey was conducted, the U.S. attorney in the case announced that the Federal Bureau of Investigation (FBI) had concluded that Bruce Ivins, a senior researcher at the U.S. Army Medical Research Institute of Infectious Diseases, was the "only person responsible for these attacks" (Johnson et al. 2008). (Ivins had committed suicide in late July [Shane and Lichtblau 2008]). The FBI has released some of the evidence they used to implicate Ivins as the suspect (Willman and Savage 2008).

[5]The term "dual use" in this context refers to legitimate research knowledge and materials that could be misused for malicious purposes; it does not refer to activities banned by the Biological and Toxin Weapons Convention (BWC) that can be cloaked by a guise of legitimacy. Thus, the dilemma is that the research is legitimate and should be conducted, even if it has the potential for misuse, as opposed to illegitimate research intended to do harm, which should not be allowed. For a discussion of the multiple uses of the term "dual use," see Atlas and Dando (2006).

[6]For ongoing discussion about the dual use dilemma, see the minutes of meetings of the National Science Advisory Board for Biosecurity (NSABB) available at http://oba.od.nih.gov/biosecurity/.

ing globalization of biotechnology has meant, however, that the United States does not have a monopoly on advanced research and technologies in the life sciences (NRC 2006a). Any development of systematic policies and practices to protect against the misuse of knowledge generated by researchers in the life sciences ultimately will have to extend globally.[7] Unilateral U.S. policies could even cause harm if they disrupted international collaborations essential to the advancement of the biomedical and other life sciences research. A number of recent NRC reports have argued that maintaining U.S. national security depends on continued promotion of the open exchange of research (NRC 2007a,b, 2009). Overall, finding the right mix of policies to advance research and share knowledge in the life sciences and also to address the potential for misuse of the knowledge generated by such research present a daunting challenge for the scientific community and policy makers.[8]

The life sciences are not the only area of science to have experienced concerns about misuse of the knowledge generated from research. In the 1980s, there was great concern about the potential for information in the open scientific literature being expropriated by enemies of the United States for arms development. Much of this concern centered on the physical sciences, with nuclear weapons development the initial focus. A National Academy of Sciences report, *Scientific Communication and National Security*, issued in 1982 and commonly known as the Corson report after its chair Dale Corson, ignored the life sciences when it considered the national security risks associated with research in the United States (NAS/NAE/IOM 1982). The report underscored the importance of maintaining the openness of fundamental research regardless of the field of science, and called upon the U.S. government to keep secret only very narrowly defined scientific knowledge. It pointed to the need to use classification as the means of protecting information that could readily be misused, thereby rejecting the idea of government control of unclassified scientific data.

However, the issue of restricting unclassified information continued to arise. In 1984, the American Association for the Advancement of Science (AAAS) passed a resolution opposing "continuing governmental efforts to

---

[7]A report from the National Research Council, *Biotechnology Research in an Age of Terrorism* (NRC 2004a), recommended that the United States undertake measures to address the potential risks from dual use research. The report pointed out that this was a first step and also called for international action to prevent the misuse of science that would be adapted to address local needs and conditions as well as global concerns.

[8]See *Making the Nation Safer: The Role of Science and Technology in Countering Terrorism* (NAS/NAE/IOM 2002) for a discussion of research areas needed to enhance security against bioterrorism and *Globalization, Biosecurity, and the Future of the Life Sciences* (NRC 2006a) for a discussion of concerns about the need for enhancing biosecurity in a globalized world.

restrict the communication or publication of unclassified research" (Borras 1984). This statement reaffirmed a AAAS resolution passed in 1982, which voiced opposition to "governmental restrictions on the dissemination, exchange, or availability of unclassified knowledge" (Borras 1982). According to the AAAS Committee on Scientific Freedom and Responsibility, the second resolution was prompted by what the AAAS considered excessive executive branch efforts to prevent the export of U.S. technology to Soviet bloc countries. These efforts included requesting prepublication review of unclassified technical papers, inhibiting communication of unclassified scientific research in university classrooms and research laboratories, limiting foreign students' access to university research projects and results, censoring technical papers at professional society meetings, and restricting otherwise unclassified meetings to U.S. citizens.

In response to the Corson report and continuing concerns about the openness of science, President Reagan issued National Security Decision Directive (NSDD) 189 in 1985, which stated that to the maximum extent possible, fundamental research results should remain unrestricted, and that the appropriate mechanism for controlling information produced by federally funded research is the classification process.[9] Nevertheless, by the 1990s the issue again arose—this time for the field of cryptography. Recognizing the threat to national security that could arise from advances in cryptography, cryptographers agreed to submit articles on a voluntary basis for government review simultaneously with submission to journals and to consider requests to restrict publication of details that could harm national security (Dam and Lin 1996:417; Diffie 1996:2).

[9]NSDD-189 states: "'Fundamental research' means basic and applied research in science and engineering, the results of which ordinarily are published and shared broadly within the scientific community, as distinguished from proprietary research and from industrial development, design, production, and product utilization, the results of which ordinarily are restricted for proprietary or national security reasons. . . . It is the policy of this Administration that, to the maximum extent possible, the products of fundamental research remain unrestricted. It is also the policy of this Administration that, where the national security requires control, the mechanism for control of information generated during federally funded fundamental research in science, technology and engineering at colleges, universities and laboratories is classification. Each federal government agency is responsible for: a) determining whether classification is appropriate prior to the award of a research grant, contract, or cooperative agreement and, if so, controlling the research results through standard classification procedures; b) periodically reviewing all research grants, contracts, or cooperative agreements for potential classification. No restrictions may be placed upon the conduct or reporting of federally funded fundamental research that has not received national security classification, except as provided in applicable U.S. Statutes" (White House 1985).

EFFORTS TO ENHANCE BIOSECURITY

Following the U.S. anthrax attacks in 2001, discussions within the government and the life sciences community began to consider whether there was SBU information in the life sciences that posed a national security risk and if so what should be done to mitigate the potential for misuse. Then National Security Advisor Condoleezza Rice reaffirmed the government policy embodied in NSDD-189 in November 2001.[10] The Office of Science and Technology Policy and the Office of Management and Budget consulted with the scientific and academic communities in several meetings conducted in 2002 about potential policies to define and constrain unclassified information in the life sciences that could be used for bioterrorism (Check 2002a, b). Some openly asked whether there were areas of research in the life sciences that should be prohibited and/or specific scientific information that should not be communicated; for example, the bioethicist Arthur Caplan is quoted as having said: "We have to get away from the ethos that knowledge is good, knowledge should be publicly available, that information will liberate us. Information will kill us in the techno-terrorist age, and I think it's nuts to put that stuff on websites" (Lichtblau 2001). Such comments raised the inevitable debate as to whether scientific knowledge is value free and thus without bounds, or whether there is "dangerous research" that should not be done and "dangerous scientific information" that should not be freely communicated.[11] Indeed, the question arose as to whether the life sciences, with their fundamental presumption of openness, needed to be redefined and restructured because of potentially "forbidden knowledge" (Kempner 2005).

---

[10]The letter to former Secretary of Defense Harold Brown stated: "The key to maintaining U.S. technological preeminence is to encourage open and collaborative basic research. The linkage between the free exchange of ideas and scientific innovation, prosperity, and U.S. national security is undeniable. This linkage is especially true as our armed forces depend less and less on internal research and development for the innovations they need to maintain the military superiority of the United States. In the context of broad-based review of our technology transfer controls that will begin this year, this Administration will review and update as appropriate the export control policies that affect basic research in the United States. In the interim, the policy on the transfer of scientific, technical, and engineering information set forth in NSDD-189 shall remain in effect, and we will ensure that this policy is followed" (Rice 2001).

[11]For further discussions see Salyers (2002), Vastag (2002), Zilinskas and Tucker (2002), and Petro (2007).

Journal Editors and Authors Group

During the year following the anthrax attacks of 2001, the question of openness of scientific communication in the life sciences became a contentious topic. Some members of the scientific community viewed the possibility that information might be withheld for national security purposes as self-censorship. Others considered such actions as responsible citizenship. The controversy escalated, perhaps because of the lack of guidance as to what should constitute "forbidden knowledge" and how potentially dangerous information should be constrained (Kempner 2005; Campbell 2006; Wimmer 2006; Selgelid 2007).

In 2002 the American Society for Microbiology (ASM) requested that the National Academies consider whether national security concerns should result in a fundamental change in the communication of science, namely, whether critical details that could permit the misuse of knowledge in the life sciences should be withheld from open publications. The National Academies and the Center for Strategic and International Studies (CSIS) convened a 1-day workshop in January 2003 to discuss issues regarding scientific publication and security.[12] The challenge presented at the workshop was how to minimize the risk of bioterrorism without jeopardizing the ability to repeat experiments and validate scientific claims, which was too important to scientific advancement to fundamentally change the way scientific research is communicated (Atlas 2003).

The following day, a group of scientists, journal editors, and security experts met and drafted a "Statement on Scientific Publication and Security," the crux of which was that many of the leading journals in the life sciences would accept responsibility for screening manuscripts to reduce the risk of misuse of scientific information (see Box 1-1). The statement was simultaneously published in *Science*, *Nature*, the *Proceedings of the National Academy of Sciences* (*PNAS*), and by the ASM (Fox 2003; Journal Editors and Authors Group 2003a,b,c). The overarching principle accepted by the Journal Editors and Authors Group stated that: "there is information that, although we cannot now capture it with lists or definitions, presents enough risk of use by terrorists that it should not be published" (Journal Editors and Authors Group 2003c). The group indicated that if "the potential harm of publication outweighs the potential societal benefits," a manuscript may be rejected (Journal Editors and Authors Group 2003a,b,c). Of note, the statement by the Journal Editors and Authors Group says that publications are not the only place where science is communicated and that all scientists are responsible for moni-

[12]See http://www7.nationalacademies.org/DSC/Scientific_Openness_Homepage.html.

BOX 1-1
Statement of Principles by Journal Editors and Authors Group

First: The scientific information published in peer-reviewed research journals carries special status, and confers unique responsibilities on editors and authors. We must protect the integrity of the scientific process by publishing manuscripts of high quality, in sufficient detail to permit reproducibility. Without independent verification—a requirement for scientific progress—we can neither advance biomedical research nor provide the knowledge base for building strong biodefense systems.

Second: We recognize that the prospect of bioterrorism has raised legitimate concerns about the potential abuse of published information, but also recognize that research in the very same fields will be critical to society in meeting the challenges of defense. We are committed to dealing responsibly and effectively with safety and security issues that may be raised by papers submitted for publication, and to increasing our capacity to identify such issues as they arise.

Third: Scientists and their journals should consider the appropriate level and design of processes to accomplish effective review of papers that raise such security issues. Journals in disciplines that have attracted numbers of such papers have already devised procedures that might be employed as models in considering process design. Some of us represent some of those journals; others among us are committed to the timely implementation of such processes, about which we will notify our readers and authors.

Fourth: We recognize that on occasions an editor may conclude that the potential harm of publication outweighs the potential societal benefits. Under such circumstances, the paper should be modified, or not be published. Scientific information is also communicated by other means: seminars, meetings, electronic posting, etc. Journals and scientific societies can play an important role in encouraging investigators to communicate results of research in ways that maximize public benefits and minimize risks of misuse.

SOURCE: Journal Editors and Authors Group (2003a,b,c) and Fox (2003).

toring their communication to maximize the benefits and minimize the risks of their research.

Several journals adopted formal policies to consider dual use and the potential for misuse of the information in the manuscript during the review. Today, the Nature Publishing Group, *PNAS*, the ASM journals, and *Science* have policies in place on biosecurity. Though these policies are not uniform, they signify the concern regarding science and security among high-impact journals. At a session on ethics in publishing held at the 2008 AAAS meeting, Donald Kennedy suggested that security issues

were likely to intrude upon the peer-review process in a way that leaves editors with little control (Timmer 2008).

In practice, the authors of several contentious manuscripts reportedly considered the dual use and societal implications of their research before submitting their manuscripts for publication; several also have said that their research had been modified because of dual use concerns. Prior to the statements by the Journal Editors and Authors Group, Ron Jackson and his colleagues brought the potential dual use issues of their work on IL-4 insertion into mousepox virus to the attention of the Australian government officials before going ahead with the publication (Cohen 2002; Federation of American Scientists 2008a). Following publication and the ensuing controversy, the research was terminated because of concerns about its dual use potential (Federation of American Scientists 2008a). Eckard Wimmer has said that the original version of the manuscript on the synthesis of poliovirus (Cello et al. 2002) included a section on social and security implications of the experiment, but that the space limitations in the journal forced the removal of that section prior to publication (Federation of American Scientists 2008b; private communication from Eckard Wimmer to Kathleen Vogel). Prior to publication of his work on the synthesis of a bacteriophage (Smith et al. 2003), Craig Venter had discussions with government officials about the dual use implications of the research (Venter 2007). The authors were fully aware of the controversy that had occurred following publication of the Wimmer poliovirus synthesis paper and Venter has said he was prepared to modify the manuscript. A formal governmental review, however, did not find any reason to make modifications because of dual use concerns.

More recently, the synthesis of the 1918 influenza virus (Taubenberger et al. 2005; Tumpey et al. 2005) also received prepublication review for dual use concerns by government officials and by the NSABB (CDC 2005). The NSABB endorsed publication of the paper and did not request that any information be withheld. The board did request that a statement be added to the manuscripts stating that the work was important for public health and was conducted safely, and suggested that *Science* should include an editorial to support the case for the importance of the research (Kennedy 2005; Sharp 2005). The authors argued that there can be no absolute guarantee that the research information in the publication will not be misused: "We are aware that all technological advances could be misused. . . . But what we are trying to understand is what happened in nature and how to prevent another pandemic. In this case, nature is the bioterrorist" (*Nature* 2005).

Finally, the publication of botulinum toxin as a threat to the milk supply (Wein and Liu 2005) was reviewed for dual use concerns prior to publication:

> The *PNAS* review considered both the above criteria and a more general sense that our publication of an article must not constitute a "roadmap for terrorists" by providing anyone who intends to do harm with key information that is otherwise difficult to obtain. . . .
>
> NAS and *PNAS* representatives met with government representatives to discuss their specific concerns about the Wein and Liu article on June 7. Following this meeting, the Council of the National Academy of Sciences decided to publish the article as originally accepted (after a standard round of final copyediting), accompanied by this editorial to make clear our reasons for doing so (Alberts 2005).

Only in a few of these cases was there any discussion of modifications to the manuscripts to remove information of dual use concern. Decisions to provide accompanying editorials to explain and endorse the scientific importance of the research reflected the perceived need to make the strongest case possible for the value of such potentially controversial research.[13]

The Fink Committee

Beyond the *ad hoc* actions by the Journal Editors and Authors Group and individual journals, a continuing need remained for broader consensus building within the life sciences community regarding issues of dual use research and the potential misuse of the life sciences for bioterrorism. In June 2001, the National Academies began planning a project that led to the formation of the Committee on Improving Research Standards and Practices to Prevent Destructive Application of Advanced Biotechnology. The committee, which began work in spring 2002, was charged with examining how to address the potential risks posed by dual use research in the United States. The committee addressed the critical question of how to define what is dangerous and how to design a system that could minimize that danger while allowing legitimate biomedical research to proceed.

Released in October 2003, the committee's report, *Biotechnology Research in an Age of Terrorism* (also known as the Fink report after the chair of the committee, Gerald Fink of Massachusetts Institute of Technology [MIT]), was published as a book a few months later (NRC 2004a). Box 1-2 contains a summary of the report's major recommendations.

[13]The only statistic available regarding changes in research communication in response to dual use research of concern is that, of the 16,000-plus manuscripts reviewed by the 11 journals of ASM since it began screening manuscripts for dual use research, only 4 manuscripts have actually been modified in any way (these were published with minor modifications) and only 2 others were not published because the authors were unwilling to provide full methodological details, which the ASM regards as essential (Kaplan 2008).

Using several published studies as examples of "contentious research" (Epstein 2001), the committee described the dual use dilemma, which it defined as occurring because the "same technologies can be used legitimately for human betterment and misused for bioterrorism" (NRC

BOX 1-2
Key Recommendations: *Biotechnology Research in an Age of Terrorism*

Recommendation: Educating the Scientific Community
We recommend that national and international professional societies and related organizations and institutions create programs to educate scientists about the nature of the dual use dilemma in biotechnology and their responsibilities to mitigate its risks.

Recommendation: Review at the Publication Stage
We recommend relying on self-governance by scientists and scientific journals to review publications for their potential national security risks.

Recommendation: Harmonized International Oversight
We recommend that the international policymaking and scientific communities create an International Forum on Biosecurity to develop and promote harmonized national, regional, and international measures that will provide a counterpart to the system we recommend for the United States.

Recommendation: Review of Plans for Experiments
We recommend that the Department of Health and Human Services (DHHS) augment the already established system for review of experiments involving recombinant DNA conducted by the National Institutes of Health to create a review system for seven classes of experiments (the Experiments of Concern) involving microbial agents that raise concerns about their potential for misuse.

Recommendation: Creation of a National Science Advisory Board for Biodefense
We recommend that the Department of Health and Human Services create a National Science Advisory Board for Biodefense (NSABB) to provide advice, guidance, and leadership for the system of review and oversight we are proposing.

Recommendation: Additional Elements for Protection Against Misuse
We recommend that the federal government rely on the implementation of current legislation and regulation, with periodic review by the NSABB, to provide protection of biological materials and supervision of personnel working with these materials.

Recommendation: A Role for the Life Sciences in Efforts to Prevent Bioterrorism and Biowarfare
We recommend that the national security and law enforcement communities develop new channels of sustained communication with the life sciences community about how to mitigate the risks of bioterrorism.

SOURCE: National Research Council (2004a).

2004a:1). The Fink committee recommended a bottom-up approach to reduce the threat of misuse of life sciences research by mobilizing the scientific community to police itself. However, it also envisioned a role for the federal government analogous to that played by the Recombinant DNA Advisory Committee (RAC) of the NIH[14] and recommended the creation of a National Science Advisory Board for Biodefense. The report proposed a system that would establish stages from proposal to publication at which scientists would review experiments and their results to provide public assurance that advances in biotechnology with potential applications for bioterrorism or biological weapons development were receiving responsible oversight. Where conducting research posed an immediate risk (i.e., the presence of imminent danger), the committee recommended classification as the appropriate method of protecting the public and national security. The fundamental conclusion of the committee, however, was that while some information could be dangerous, most is not, and self-governance by the scientific community should be relied on to reduce the potential misuse of legitimate science. The proposed system thus sought to protect scientific enquiry from untoward government interference and to permit open communication to the maximum extent possible for the public good.

The committee's report listed seven classes of experiments it believed "illustrate the types of endeavors or discoveries that will require review and discussion by informed members of the scientific and medical community before they are undertaken or, if carried out, before they are published in full detail" (NRC 2004a:5). The seven types of "Experiments of Concern" highlighted in the report include research that could

- Demonstrate how to render a vaccine ineffective;
- Confer resistance to therapeutically useful antibiotics or antiviral agents;
- Enhance the virulence of a pathogen or render a nonpathogen virulent;
- Increase transmissibility of a pathogen;
- Alter the host range of a pathogen;
- Enable the evasion of diagnostic/detection modalities; and
- Enable the weaponization of a biological agent or toxin (NRC 2004a:5–6).

[14]For further information, see http://www4.od.nih.gov/oba/rac/aboutrdagt.htm.

Creation of NSABB

The U.S. government adopted several of the Fink report's key recommendations, including the recommendation to create a national advisory body. In 2004, NSABB was established under the auspices of the NIH Office of Biotechnology Activities, which also oversees RAC. The advisory committee's charge included providing guidance for the development of:

- A system of institutional and federal research review that allows for fulfillment of important research objectives while addressing national security concerns;
- Guidelines for the identification and conduct of research that may require special attention and security surveillance;
- Professional codes of conduct for scientists and laboratory workers that can be adopted by professional organizations and institutions engaged in life science research;
- Materials and resources to educate the research community about effective biosecurity; and
- Strategies for fostering international collaboration for the effective oversight of dual use biological research (NSABB 2004).

The NSABB established working groups to develop draft guidance regarding:

1. "Criteria" for identification of dual use research of concern,
2. "Codes of Conduct,"
3. "Communication" of methodology and research results,
4. "Oversight" of dual use research of concern,
5. "Education and Training" of scientists and laboratory workers regarding dual use,
6. "Biosecurity issues raised by Synthetic Genomics," and
7. "International" dialogue on the broad topic of dual use research of concern.[15]

A number of the NSABB working groups have produced reports or proposals or undertaken activities. Recognizing that a wide variety of life sciences research could somehow be considered dual use, the NSABB has attempted to narrow its focus to "dual use research *of concern*" (DURC), which it defines as:

---

[15]Information about all these activities and related documents may be found on the NSABB Web site at http://oba.od.nih.gov/biosecurity/biosecurity_documents.html.

> research that, based on current understanding, can be reasonably anticipated to provide knowledge, products, or technologies that could be directly misapplied by others to pose a threat to public health and safety, agricultural crops and other plants, animals, the environment, or materiel (NSABB 2007:17).

The NSABB has developed criteria for determining which experiments could be considered DURC (see Box 1-3).

The NSABB has also recommended strategies for communicating research at all stages, from proposal to publication. The NSABB education committee is considering methods to educate scientists about DURC, and its international committee has been engaging intergovernmental organizations, international scientific organizations, individual scientists, journal editors, and policy makers from other nations on oversight of dual use

BOX 1-3
NSABB Criteria for Dual Use Research of Concern

1. Is it likely that the research could:
   a. Enhance the harmful consequences of a biological agent or toxin
   b. Disrupt immunity or the effectiveness of an immunization without clinical and/or agricultural justification
   c. Confer to a biological agent or toxin, resistance to clinically and/or agriculturally useful prophylactic or therapeutic interventions against that agent or toxin or facilitate their ability to evade detection methodologies
   d. Increase the stability, transmissibility, or the ability to disseminate a biological agent or toxin
   e. Alter the host range or tropism of a biological agent or toxin
   f. Enhance the susceptibility of a host population
   g. Generate a novel pathogenic agent or toxin or reconstitute an eradicated or extinct biological agent

2. Additional Review Assessment
   a. The potential for, and the ways in which, information, technologies, or biological agents from the research could be misused to pose a threat to public health and safety, agricultural crops and other plants, animals, the environment, or materiel
   b. The likelihood that the information might be misused
   c. The potential impacts of misuse
   d. Strategies for mitigating the risks that information from the research could be misused

SOURCE: NSABB (2007).

research.[16] As discussed further below, in June 2007 the NSABB released a *Proposed Framework for the Oversight of Dual Use Life Sciences Research*, which it hopes will lead to common oversight standards across all federal agencies funding such research (NSABB 2007). A public informational meeting held at NIH in July 2008 allowed members of the scientific and academic communities to voice concerns about the proposed oversight framework; it appears that refinements will be necessary to satisfy those concerns.

Note that, while many in the biosecurity community would agree that practicing life scientists should be educated about the risks of misuse of biotechnology and that there is value in building norms and a taboo against biological weapons, there are real philosophical differences regarding the value of a Hippocratic-type oath for scientists. Proponents believe that swearing an oath against the misuse of the life sciences would contribute to the creation of an effective "web of prevention" against biological weapons (Revill and Dando 2006). Some, however, feel that reciting an oath brings no added value: "A Hippocratic Oath for life scientists would be nice window dressing, but it would simply be that. The more important method to ensure that graduating life scientists are ethical and responsible citizens is the oversight provided by their laboratory supervisors, mentors, and/or thesis advisers" (Kahn 2007). Furthermore, skeptics point out that physicians who have taken the Hippocratic oath have put that oath aside for nationalistic reasons to participate in biological weapons programs; for example Dr. Kanatjan Alibekov, a physician and scientist who directed biological weapons research and development programs in the former Soviet Union, considered the security interests of the Soviet Union more important than the Hippocratic oath he had taken (Alibek and Handelman 1999).

International Efforts

As already noted, biotechnology research is a genuinely global enterprise (NRC 2006a). The NSABB and others in the United States recognize that efforts regarding dual use research will ultimately need to be extended to create a global "web of prevention" to reduce risks of the misuse of life sciences research and knowledge.[17]

---

[16]Information about all these activities and related documents can be found on the NSABB Web site at http://oba.od.nih.gov/biosecurity/biosecurity_documents.html.

[17]In 2002, the International Committee of the Red Cross launched an initiative on Biotechnology, Weapons, and Humanity, calling for a "web of prevention" to address the risks that technologies from the life sciences could be used for hostile purposes. In addition to a number of proposals for national and international legal measures to support implementation of the BWC, the initiative recommended including education about the risks of

The fundamental commitments not to use disease as a weapon are embodied in the Geneva Protocol, which was signed in 1925 and entered into force in 1928, and the BWC, which was signed in 1972 and entered into force in 1975.[18] Article I of the BWC states:

> Each State Party to this Convention undertakes never in any circumstances to develop, produce, stockpile or otherwise acquire or retain:
>
> (1) Microbial or other biological agents, or toxins whatever their origin or method of production, of types and in quantities that have no justification for prophylactic, protective or other peaceful purposes;
>
> (2) Weapons, equipment or means of delivery designed to use such agents or toxins for hostile purposes or in armed conflict.

As Ambassador Masood Khan of Pakistan, president of the treaty's sixth review conference, stated:

> The BWC has had marked success in defining a clear and unambiguous global norm, completely prohibiting the acquisition and use of biological and toxin weapons under any circumstances. The preamble to the Convention so forcefully states: the use of disease as a weapon would be "repugnant to the conscience of mankind." It captures the solemn undertaking of the states parties "never in any circumstances to develop, produce, stockpile or otherwise acquire or retain" such weapons. With 155 states parties, the treaty is not universal, but no country dares argue that biological weapons can ever have a legitimate role in national defense. Such is the force of the treaty (Khan 2006).

UN Security Council Resolution 1540, passed in 2004, adds a further binding international commitment against support for nonstate actors seeking to acquire weapons of mass destruction or the means of their delivery.[19] In addition, in response to a report from UN Secretary General

---

misuse as part of overall ethical training for life scientists. More information can be found at http://www.icrc.ch/Web/eng/siteeng0.nsf/html/bwh!Open. (Earlier, Graham Pearson had coined the phrase "web of deterrence," but he did not address dual use research issues [Pearson 1993]).

[18]The Geneva Protocol's formal title is the Protocol for the Prohibition of the Use in War of Asphyxiating, Poisonous or Other Gases, and of Bacteriological Methods of Warfare, and the BWC's formal title is the Convention on the Prohibition of the Development, Production and Stockpiling of Bacteriological (Biological) and Toxin Weapons and on Their Destruction.

[19]"All states have three primary obligations under UNSCR 1540 relating to such items: to prohibit support to non-State actors seeking such items; to adopt and enforce effective laws prohibiting the proliferation of such items to non-State actors, and prohibiting assisting or financing such proliferation; and to take and enforce effective measures to control these items, in order to prevent their proliferation, as well as to control the provision of funds and services that contribute to proliferation. If implemented successfully, each state's actions will significantly strengthen the international standards relating to the export of sensitive items

Kofi Annan (Annan 2006), in September 2006 the UN General Assembly passed a resolution creating a *Global Counter-Terrorism Strategy*, including a proposal to bring together "the major biotechnology stakeholders, including industry, the scientific community, civil society and Governments, into a common programme aimed at ensuring that biotechnology advances are not used for terrorist or other criminal purposes but for the public good" (United Nations 2006). The transition to a new secretary general slowed progress on convening stakeholders, but development of the idea has continued and will be launched in 2009 (Orr 2008).

Over the years, the BWC has provided the international forum for discussions of issues related to biological weapons, including dual use research. As part of its intersessional process of meetings of experts and states parties, the BWC has addressed a number of issues related to dual use research. Codes of conduct as a means to foster a culture of responsibility among scientists against creating biological weapons (Dando and Rappert 2005; Dando and Revill 2005; Pearson 2004) and for reducing the risks of misuse of dual use research (McLeish and Nightingale 2005) were important considerations of the 2005 BWC Meeting of Experts and Meeting of States Parties.[20] The states parties discussed not only codes of conduct for scientists, but also relevant codes by sector, including government, academia, and professional societies; weaknesses in relying on just codes of conduct to prevent creation of biological weapons; and relevant national policies or legislation that overlap with codes of responsible conduct. The 2008 BWC Meetings of Experts and States Parties expanded the discussion to include "oversight; education; awareness raising; and adoption and/or development of codes of conduct."[21]

Other international organizations have become engaged in dual use issues as well. In 2005 the World Health Organization (WHO) released a background paper, "Life Science Research: Opportunities and Risks for Public Health," as an initial step toward increasing engagement in the issue (WHO 2005). WHO then held a workshop in October 2006 on "Life Science Research and Global Health Security." The workshop report recommends the creation of a standing scientific advisory group to advise the WHO director-general on biosecurity, including both improved biosafety and responsible oversight of research (WHO 2007). WHO is also under-

---

and support for proliferators (including financing) and ensure that non-State actors, including terrorist and black-market networks, do not gain access to chemical, nuclear or biological weapons, their means of delivery or related materials" (U.S. Department of State 2006).

[20]For details, see http://disarmament.un.org/wmd/bwc/annualmeetings/listofdocs-2005%20expert%20mtgs.html.

[21]Available at http://www.unog.ch/80256EE600585943/(httpPages)/F1CD974A1FDE4794C125731A0037D96D?OpenDocument.

taking a number of collaborative activities, including regional workshops addressing both biosafety and biosecurity issues.

The Organisation for Economic Co-operation and Development (OECD) Global Futures Programme created a Web site (*www.biosecurity-codes.org*) to provide information about national and international activities. Another source of more general information about national laws and policies is the BWC National Implementation database, created as part of the 2003 intersessional process and updated and enhanced for the 2007 discussions.[22]

The involvement of organizations such as WHO and OECD has added the important elements of global health and economic development to the more traditional security concerns represented by the BWC. This has also served to emphasize the need for a mix of policies to ensure that efforts to reduce the risk of misuse also allow continued scientific progress.

A number of international scientific organizations have also become engaged on biosecurity issues. In 2004 the InterAcademy Panel on International Issues (IAP), a network of 100 of the world's academies of science, created a Biosecurity Working Group, and issued a Statement on Biosecurity in December 2005 that offers principles to guide any scientific body that wants to develop a code of conduct to protect against the misuse of information in the life sciences.[23] The IAP, through its Working Group on Biosecurity, is continuing its work on education and awareness-raising efforts including two international forums in collaboration with other international scientific organizations.[24] AAAS addressed the 2008 intercessional meeting of states parties on development and implementation of education programs on dual use research discussing their recent workshop documenting existing education programs on dual use research, discussing the design of similar education programs, and highlighting challenges for implementing such programs.[25] Another nonprofit organization with a particular emphasis on engaging industry, the International Council for the Life Sciences (ICLS), is seeking to enhance global

---

[22]The database can be found at http://www.unog.ch/unog/website/disarmament.nsf/(httpPages)/A68F6976314FD5FDC1257478004F304F?OpenDocument&unid=74C574F0A6EEC10EC12572BC00334CC5.

[23]The IAP Biosecurity Working Group has members from the academies of China, Cuba, the Netherlands (chair), Nigeria, the United Kingdom, and the United States. The statement can be found at http://www.interacademies.net/Object.File/Master/5/399/Biosecurity%20St.pdf.

[24]See http://www7.nationalacademies.org/biosecurity/International%20Biosecurity%20Project.html for further information about both meetings. A summary report of the 2nd International Forum was published by the National Research Council (NRC 2008).

[25]The AAAS workshop was held on November 12, 2008, in Washington, D.C. A summary of the workshop and resulting recommendations can be found at http://cstsp.aaas.org/files/AAAS_workshop_report_education_of_dual_use_life_science_research.pdf.

biological security and safety through international standards and sharing of best practices.[26]

A number of national academies have also undertaken activities. For example, in part because of its work with the IAP, the Royal Netherlands Academy of Arts and Sciences developed a code of conduct for Dutch scientists at the request of the government of the Netherlands.[27] The Royal Society has published a summary of its activities, which include a number of meetings and policy papers (Royal Society 2008).

Another important national effort with broader implications has come from the United Kingdom. The three largest UK funders of biomedical research announced in September 2005 that they would ask applicants to indicate whether their proposed research had dual use potential and to include dual use considerations in reviews. The joint policy statement from the Biotechnology and Biological Sciences Research Council, the Medical Research Council, and the Wellcome Trust (BBSRC/MRC/Wellcome Trust 2005) identifies a series of agreed actions that the three organizations have implemented to raise awareness and to help ensure that any risks of misuse associated with research proposals are considered at the grant application stage.

The NSABB's international committee has conducted three roundtable dialogues in partnership with WHO. The first was held in February 2007 with scientists and policy makers from 17 nations and several intergovernmental organizations, the second was held in October 2007 with members of U.S. and international professional scientific organizations and societies, and the third was held in November 2008 with 130 participants from approximately 40 countries as well as international organizations and scientific bodies.[28]

The participants in the first dialogue highlighted the importance of life science research and the need to protect advancement in the life sciences while minimizing any risks to international or national security. They agreed that dual use research needs to be clearly defined and that there should be guidance for dealing with any problems or challenges that might arise while identifying or addressing dual use potential. The participants believed that the scientific community should be broadly educated about the dual use dilemma and should adopt a culture of responsibility for minimizing security risks. Finally, they highlighted the

---

[26]More information may be found on the ICLS Web site at http://www.iclscharter.org/home.html.

[27]The code can be found at http://www.knaw.nl/publicaties/pdf/20071092.pdf.

[28]For more information, see http://oba.od.nih.gov/biosecurity/biosecurity_documents.html.

global nature of the life sciences and supported global efforts to deal with the dual use dilemma.

The second roundtable discussion built upon these concepts by engaging American and international scientific organizations to discuss their current efforts in dual use research and help develop an international engagement strategy for effectively reaching out to the scientific community. Some key challenges highlighted at this meeting were problems with defining dual use research, biosafety, and biosecurity; determining how to measure the success of national programs for oversight of dual use research; and establishing common guidance and standards regarding dual use research.

The third roundtable included breakout sessions to facilitate exploring key topics—awareness raising, training, and education; culture of responsibility and codes of conduct; review of research proposals and guidelines for review; and scientific presentations, publications, and communication—in depth in order to develop an inventory of current and planned activities, share experiences with development and implementation, and make recommendations for further work. With regard to education, for example, participants recommended the development of standard components of educational programs and leveraging current educational efforts in areas such as biosafety, ethics, and responsible conduct of research, while recognizing the need to adapt approaches to specific needs and educational contexts. Participants recommended review processes across the life cycle of research, although with an emphasis on reviews prior to the time a manuscript is submitted for publication. More broadly, participants saw important opportunities to achieve policies and practices that could both sustain continued progress in the life sciences and mitigate the risks associated with dual use research.

SPECTRUM OF APPROACHES TO MANAGE RISKS OF DUAL USE RESEARCH

As discussions about the dual use dilemma have moved forward, a variety of proposed approaches for safeguarding materials, research, and knowledge in the life sciences has emerged.[29] Mandatory government requirements (i.e., laws and regulations) are at one end of the spectrum while voluntary responsible conduct within the life sciences community

[29]Dual use research, as defined by the NSABB, deals with the knowledge, tools, and techniques produced by research rather than biological materials. The implementation of the select agent regulations, which affects aspects of how research is conducted, has tended to foster policy discussions in which dual use research and select agents are conflated. The survey described in this report asked questions about both select agents and dual use research.

(i.e., self-governance) lies at the other end of the spectrum.[30] Other proposals combine regulatory and voluntary measures to achieve a mix of top-down and bottom-up approaches.

Regulations on Possession of Potentially Dangerous Pathogens and the American Life Sciences Community

On November 25, 1969, President Nixon issued National Security Decision Memorandum 35, which renounced the "use of lethal methods of bacteriological/biological warfare. The United States bacteriological/ biological programs will be confined to research and development for defensive purposes (immunization, safety measures, et cetera)" (White House 1969:2–3).[31] This order did not regulate possession of potentially dangerous pathogens, but instead focused on the purpose of the research. The BWC took a similar approach; as already noted, Article 1 states that "Each State Party to this Convention undertakes never in any circumstances to develop, produce, stockpile or otherwise acquire or retain: (1) Microbial or other biological agents, or toxins whatever their origin or method of production, of types and in quantities that have no justification for prophylactic, protective or other peaceful purposes."

The BWC calls on states parties to enact legislation to support implementation of the convention. The United States passed the Biological Weapons Anti-Terrorism Act of 1989 (P.L. 101-298, May 22, 1990), which established penalties for violating the Convention's prohibitions, unless "(1) such biological agent, toxin, or delivery system is for a prophylactic, protective, or other peaceful purpose; and (2) such biological agent, toxin, or delivery system, is of a type and quantity reasonable for that purpose." Since President Nixon had already ended the U.S. offensive biological weapons program two decades earlier, the U.S. scientific community took very little notice of this new law.

Although biological agents, including pathogenic microorganisms, are not weapons banned by the Geneva Convention or the BWC, there has been continuing concern that a variety of agents could be used as weap-

---

[30]"Ethics and law are related, but they are not the same. Law draws the line we cannot cross without becoming 'outlaws.' Even if we do not like it, we must nonetheless follow it (while working to change it) or risk . . . being prosecuted for being an outlaw. . . . Americans can go to jail for violating the law, but not for violating codes of ethics. We aspire to uphold ethics—we deserve praise (at least some) for behaving 'ethically'; whereas we deserve none for simply following the law, some of which is in fact made up of 'legal technicalities'" (Annas 2006).

[31]The Memorandum stated that "This does not preclude research into those offensive aspects of bacteriological/biological agents necessary to determine what defensive measures are required" (White House 1969:3).

ons. The first legislation that included provisions dealing directly with possession of biological agents that were highly dangerous was the Antiterrorism and Effective Death Penalty Act of 1996 (P.L. 104-132, April 24, 1996), which was enacted after a U.S. scientist attempted to obtain plague-causing bacteria for potentially nefarious purposes (Gronvall 2008). The act established the "select agent" program to control the transfer of certain biological agents that were considered especially dangerous.[32] The program was created to prevent bioterrorism and protect public safety while not unduly hindering research using select agents.

After the anthrax attacks of 2001, the United States sought to establish a rigorous formal oversight system to decide who could possess microorganisms and toxins that could be used as weapons. Appropriately defined, such a system would ensure that pathogens and toxins would be available to legitimate researchers and denied to those who should not possess dangerous agents. The aim of material control efforts for the life sciences soon focused on ways to ensure that any individual given access to select agents was trustworthy and that these agents would be secure within each facility that housed dangerous pathogens and toxins.[33]

The USA PATRIOT Act of 2001 (P.L. 107–56, October 26, 2001) established restrictions on the possession of select agents by aliens from countries designated as supporting terrorism and for other individuals who were deemed unsafe for specific objective reasons, such as those ineligible to purchase handguns. The act also made it an offense for a person to knowingly possess any biological agent, toxin, or delivery system of a type or in a quantity that, under the circumstances, is not reasonably justified by prophylactic, protective, bona fide research or other peaceful purpose. The provisions of the Patriot Act were subsequently augmented by the Public Health Security and Bioterrorism Preparedness and Response Act, known as the Bioterrorism Preparedness Act of 2002 (P.L. 107–188,

---

[32]Since its inception the select agent program has been modified to include oversight of possession as well as transfer of certain agents; the list of agents that might be used as biological weapons to attack humans has also increased, and potential agricultural biothreat agents have been added. More information on the program can be found at http://www.cdc.gov/od/sap/.

[33]In 2008, after the FBI identified a researcher at the USAMRIID as the perpetrator of the 2001 anthrax attacks, the issue of whom to trust and how to ensure the reliability of those with access to dangerous pathogens again gained special attention. The NSABB was charged with developing a system for determining the reliability of researchers with access to potentially dangerous pathogens and toxins. The U.S. Congress, following a report by the Commission on the Prevention of WMD Proliferation and Terrorism (2008), also began to consider taking action to enhance the protection of biological agents against acquisition by terrorists; see, for example, the hearings of the U.S. Senate Homeland and Government Affairs Committee at http://hsgac.senate.gov/public/index.cfm?Fuseaction=Hearings.Detail&HearingID=d0d0b4c1-d1d1-4b7a-9c16-fd9af22d97e0.

June 12, 2002). This act added requirements for regulations governing possession of select agents, including approval for laboratory personnel by the attorney general following a background check by the FBI.[34] While some scientists have expressed concern about the impact of the select agent regulations on the advancement of science, others have worried that additional oversight is needed. On January 9, 2009, President George W. Bush issued an executive order, *Strengthening Laboratory Biosecurity in the United States*, that established a process for new government oversight to further "the policy of the United States that facilities that possess biological select agents and toxins have appropriate security and personnel assurance practices to protect against theft, misuse, or diversion to unlawful activity of such agents and toxins" (White House 2009).

In addition to the United States, Great Britain and some other nations have also taken the approach of legally restricting who can have access to the microorganisms and toxins thought to have the highest risk of use for bioterrorism and have established regulations for the possession of these agents. In the United Kingdom, the Anti–Terrorism, Crime and Security Act of 2001 strengthened controls on access to pathogens and toxins used in research laboratories. All UK facilities handling pathogens and toxins in hazard group 2, 3, or 4 must be registered, and strict control of hazard group 3 and 4 organisms is in place.[35] In Germany, the Act on the Reform of the Communicable Diseases Law (Communicable Diseases Law Reform Act) contains provisions that limit the distribution of pathogens to authorized individuals. Although the purpose of this act is to prevent the spread of communicable diseases in human beings by early detection of infections, the provisions that restrict the distribution of human pathogens are intended to contribute to the deterrence of bioterrorism.[36]

Policies implemented in the United States after September 11 have significantly increased the level of scrutiny and the time involved in the

---

[34]Clinical laboratories were granted a special exemption to permit them to legally isolate and identify (and thereby possess) select agents cultured from patients as part of the medical diagnostic process, even if they were not registered to possess select agents. This is critical for medical diagnoses where there is no way to predict what disease a patient might have, thereby precluding the ability to register for specific select agents. The clinical laboratories, however, are required to destroy any select agents or transfer them to a registered laboratory that is permitted to possess that select agent within a few days, and they must also notify public health authorities whenever a select agent has been isolated and identified.

[35]The approved list of biological agents established by an Advisory Committee on Dangerous Pathogens is available at http://www.hse.gov.uk/pubns/misc208.pdf. The categories are similar to the biosafety levels for laboratories in the United States, but the lists of agents within the categories are different.

[36]Available at http://www.unog.ch/80256EDD006B8954/(httpAssets)/E0C6A0305240C953C12574AA0044D41B/$file/BWC_MSP_2008_MX-Statement-Germany-080818-PM.pdf.

process of obtaining a visa for a short or extended stay and increased the tracking of students and scholars once they arrive on U.S. shores.[37] Although there are differing views within the scientific community regarding the consequences of the select agent regulations, several in the scientific community consider that increased restrictions on foreign students and scholars through measures such as more stringent visas pose a far greater threat to the conduct of life sciences research.[38] As the presidents of the National Academies have stated: "Our visa processing system not only must provide genuine security against those who might do us harm, but also keep our borders open to the stream of scientific and technical talent that fuels our progress . . . the U.S. scientific, engineering, and health communities cannot hope to maintain their present position of international leadership if they become isolated from the rest of the world" (Alberts et al. 2002).[39] As noted earlier, in addition to limiting visas, the restrictions in place under the Patriot and Bioterrorism Preparedness acts prohibit any national of a country currently designated by the Department of State as a state sponsor of terrorism from having access to select agents.

Oversight of Research

A more contentious issue than control of pathogens and toxins is whether there should be oversight of life science research and communication, and if so, what kind. The BWC does not include measures to limit research, although its intersessional processes and confidence-building measures do collect information about national activities and address how continuing advances in research affect the implementation and operation of the treaty. This brief review highlights some of the major proposals from government and private sources but does not attempt to do justice to the range and variety of ideas for addressing dual use research.

The Controlling Dangerous Pathogens Project of the Center for International and Security Studies at Maryland has developed a proposal for

---

[37]The National Academies created a special International Visitors Office to provide a resource on visa policy and to collect data on cases involving delays or difficulties experienced by students or visiting scientists. For further information see http://www7.nationalacademies.org/visas/.

[38]In addition to visa issues, the issue of the impact of export controls on scientific research through restrictions on the transfer of dual use equipment and materials has drawn substantial attention from the scientific community. These problems are dealt with in another NRC report, *Beyond Fortress America: National Security Controls on Science and Technology in A Globalized World* (NRC 2009).

[39]See NRC (2006b) for a discussion of this issue focused on international graduate students.

the oversight of research in the life sciences that involves enhancing the responsibilities of institutional biosafety committees (IBCs) to include review of biosecurity implications of research activities (Steinbruner et al. 2007). This review would be complemented by regional IBCs and international IBCs. The project has also developed an electronic questionnaire for these entities to help them evaluate research activities for their dual use potential. A consequence of the proposal would be that the performance and communication of specific life sciences research falling within a narrow set of parameters could be prevented at a series of review stages, including potential review by a new international organization. Van Aken (2006) has also called for internationally harmonized and legally binding rules for conducting dual use research, in order to prevent the misuse of biological knowledge.

Under the scientific self-governance approach, the Center for Biosecurity of the University of Pittsburgh Medical Center supports responsible stewardship and self-governance by life scientists without mandated governmental restrictions (Kwik et al. 2003). The basic tenets of this perspective are that scientists should be involved in determining which experiments are dangerous and made aware of the potential misuse of biotechnology; security measures should be thoughtfully developed and integrated into the existing research culture; biosecurity awareness and measures should begin with scientists rather than be prescribed by governments; periodic evaluation and improvement of biosecurity measures should occur; and through collaborations, the norms and ethical standards to prevent misuse of biotechnology should be established and propagated globally.

The Fink committee also supported a self-governance approach for protecting scientific knowledge from misuse. However, as already mentioned, it supported an advisory role for the federal government in providing guidelines along the model of RAC administered by the NIH. Although formal federal regulations or guidelines have yet to be devised, the IBCs of three of the six member institutions in the Southeast Regional Center of Excellence for Emerging Infections and Biodefense (SERCEB) have assumed responsibility for dual use review (Davidson et al. 2007).

Providing federal government guidelines and then relying upon self-governance by research institutions is the key to the proposed NSABB oversight system. As mentioned earlier, in June 2007 the NSABB released its *Proposed Framework for the Oversight of Dual Use Life Sciences Research*, which it hopes will lead to common oversight standards across all federal agencies funding such research (NSABB 2007). Under the proposed system, each principal investigator would make an initial assessment of the potential for misuse, and IBCs would expand the scope of their current reviews to include a biosecurity evaluation. The proposed framework

is currently undergoing an interagency review.[40] Some have expressed concerns that this may result in mandated actions, including: training for all life scientists receiving federal funds, signing of codes of conduct (equivalent to taking an oath), and reporting of suspicions about others' research activities. In particular, concern has been expressed about what actions will be required once awareness of dual use issues increases (Stebbins 2008).

One area of life sciences research that has attracted particular attention regarding potential oversight is synthetic biology. This growing field combines elements of biological science, chemistry, and engineering into a highly interdisciplinary area of research. Synthetic biology has the potential for construction of bioengineered microorganisms that can mass-produce drugs to treat disease, detect and break down toxic chemicals to reverse polluted sediments and water, and generate new energy forms to help solve the energy crisis. There also is substantial concern, however, regarding the potential for the creation of dual use products, either intentionally or unintentionally, that could function as biological weapons in the hands of terrorists. The NSABB has proposed greater oversight of synthetic biology (NSABB 2006). In addition, several groups have proposed governance strategies for preventing misuse of synthetic biology that combine government oversight and self-governance. In the United States this includes a collaboration among the Venter Institute, CSIS, and a scientist at MIT (Garfinkel et al. 2007a,b), as well as a group in the Goldman School of Public Policy at the University of California, Berkeley (Maurer et al. 2006). Internationally, the International Consortium for Polynucleotide Synthesis (Bügl et al. 2006) and the Industry Association for Synthetic Biology (Bernauer et al. 2008) have also made proposals.

LIFE SCIENTISTS' ATTITUDES AND AWARENESS OF DUAL USE ISSUES

Since responsible conduct is considered key to the success of self-governance measures to protect against the misuse of the life sciences, there is concern that life scientists are insufficiently aware of and engaged in discussions about biosecurity and dual use issues. Most life scientists have little experience with the issues of biological weapons. Without conscious personal effort or systematic education, very few life scientists working today in the United States or overseas would have reason to know of past offensive weapons programs. Between the end of the U.S.

[40]The status of the NSABB oversight document in the U.S. federal agency review process was discussed at the December 10, 2008, NSABB meeting. The archived Webcast can be viewed at http://oba.od.nih.gov/biosecurity/nsabb_past_meetings.html#dec2008.

biological weapons program in 1969 and the anthrax mailings in October 2001, few biologists had connections to the national security branches of government.[41] This has left the life sciences community poorly prepared to interact with the security community and divided about whether dual use issues are really of serious concern. Many studies and conferences on biosecurity thus end with recommendations for greater awareness raising and education in the life sciences community.[42] The report of the Commission on the Prevention of WMD Proliferation and Terrorism, released in 2008, gave the scientific community this ominous warning:

> The choice is stark. The life sciences community can wait until a catastrophic biological attack occurs before it steps up to its security responsibilities. Or it can act proactively in its own enlightened self interest, aware that the reaction of the political system to a major bioterrorist event would likely be extreme and even draconian, resulting in significant harm to the scientific enterprise (Commission on the Prevention of WMD Proliferation and Terrorism 2008:26).

Many of the major U.S. scientific organizations and professional societies, however, including the National Academies, delayed undertaking significant education and awareness-raising initiatives until the U.S. government clearly defined its policies. In the interim, a number of universities and organizations, such as the NIH-supported Regional Centers for Excellence, the Federation of American Scientists (FAS), and AAAS have undertaken educational activities.[43]

---

[41]In this regard, the life sciences community differs markedly from the physics and engineering communities, which have been continuously aware of security concerns associated with government-sponsored weapons research programs since at least World War II. The closest analogy in the life sciences is development and advancement of recombinant DNA technology. Other examples of social considerations to the life sciences include human subjects research (e.g., the Tuskegee syphilis experiments) and the welfare of research animals. Although the United States had a biological weapons program between 1943 and 1969, it has not used biological weapons in times of war. At the time of the Asilomar conference (1975) and the development of the BWC (1972), there was extensive discussion about the harmful uses of recombinant DNA, but the concerns of biotechnology and offensive biological weapons programs did not propagate that interest or awareness of science policy to the broader life sciences community.

[42]See, for example, NRC (2004a,b, 2006a), and the discussions at the 2008 BWC meetings of experts and states parties (http://www.unog.ch/80256EE600585943/(httpPages)/92CFF2CB73D4806DC12572BC00319612?OpenDocument).

[43]FAS *Case Studies in Dual use Biological Research* illustrate the dual use potential of actual life science research via interviews with the lead researchers. The case studies provide a historical background on bioterrorism, bioweapons, and the current laws, regulations, and treaties that apply to biodefense research. They include interviews with researchers as well as the primary scientific research papers and discussion questions meant to raise awareness about the importance of responsible biological research. The case studies are available at

In September 2005, the National Academies and AAAS cohosted a workshop, "Education and Raising Awareness: Challenges for Responsible Stewardship of Dual Use Research in the Life Sciences," that brought together over 50 participants to share information and explore ways to engage and educate the life science research community most effectively.[44] The workshop addressed (1) opportunities for outreach presented by the BWC, NSABB, and others interested in biosecurity education; (2) challenges and opportunities in framing the discourse on biosecurity as well as differences in those factors created by multiple stakeholders and audiences; (3) case studies of outreach; and (4) the roles of codes of ethics, conduct, and practices in raising awareness.[45]

Although the workshop did not produce formal recommendations, one theme that emerged from the discussions was the need to move beyond anecdotal evidence to empirical information about perceptions and attitudes among life scientists that could provide a basis for creating programs and assessing their impact. Remarkably few data are available today. There have been numerous surveys of scientists on other topics related to professional responsibility, such as ethics, responsible conduct, and financial conflicts of interest.[46] Several small-scale studies using interviews or surveys with practicing scientists in the United States and the

---

http://www.fas.org/programs/ssp/bio/educationportal.html. The SERCEB Policy, Ethics and Law Core has developed an online module to assist those involved with the biological sciences to better understand the dual use dilemma of some life science research. This module is intended for graduate students and postdoctoral fellows, faculty members, and laboratory technicians involved in biological research in microbiology, molecular genetics, immunology, pathology, and other fields related to emerging infectious diseases and biodefense. The module consists of an approximately 20-minute online presentation followed by a brief assessment. The recently updated module is available at http://www.serceb.org/modules/serceb_cores/index.php?id=3. It has been used by over 600 people. AAAS held a workshop in November 2008 to document and discuss existing education programs on dual use research and to highlight challenges and provide recommendations for development and implementation of other educational programs on dual use research. A summary of the workshop is available at http://cstsp.aaas.org/files/AAAS_workshop_report_education_of_dual_use_life_science_research.pdf.

[44]The participants included staff from several U.S. government agencies, representatives of nongovernmental organizations such as the ASM, the FAS, and the Royal Society, as well as university professors from the United States and Britain.

[45]For details, see http://www7.nationalacademies.org/dsc/Biosecurity_Workshop.html.

[46]See, for example, Blumenthal et al. (1996a,b, 1997, 2006); Eastwood et al. (1996); Wenger et al. (1997, 1999); Korenman et al. 1998; Rabino (1998, 2003a,b); Berk et al. (2000); McCrary et al. (2000); MORI (2000); Price et al. (2001); Campbell et al. (2002); Martinson et al. (2005, 2006); Union of Concerned Scientists (2005a,b, 2006, 2008); De Vries et al. (2006); Hansen et al. (2006); Royal Society/Research Councils UK/The Wellcome Trust 2006; and Vogeli et al. (2006).

United Kingdom, including those in biodefense research in the U.S. case, offer perspectives on dual use research.[47]

Seminars and discussions conducted by Brian Rappert and Malcolm Dando provide insights about the views of some 2,500 scientists in 13 countries. A series of about 100 information exchange seminars have served as focus groups to gauge awareness of dual use issues, ascertain attitudes about potential oversight mechanisms, and raise awareness of the potential misuse of the life sciences for bioweapons development. As a result, Rappert and Dando have compiled substantial qualitative information. They report that their 26 seminars in the United Kingdom "found little evidence that [scientists] regarded bioterrorism or bioweapons as a substantial threat; considered that developments in the life sciences research contributed to biothreats; were aware of the current debates and concerns about dual use research; or were familiar with the BTWC"; similar results were found in the 28 seminars they conducted in Finland, South Africa, and the Netherlands, although not completely in the United States (Rappert 2008).[48]

With regard to measures to address the potential risk of dual use research, Rappert and Dando reported that they encountered overwhelming skepticism about the value of publication restraint, but there was a sense that some contentious experiments perhaps should not have been openly reported. For instance, they found support for doing the IL-4 mousepox research and communicating the results; to whom the results were communicated was a topic of contention, however. They also stated that only a small percentage of scientists indicated knowledge of the IL-4 study. Although some saw value in community involvement in demonstrating responsibility, there was general skepticism of the oversight approach proposed by the Fink committee. Doubts were also expressed about the feasibility of pre-project review and oversight systems for dual use research (Rappert 2007).

All of these studies provide useful information and insights. However, there is no example of a large-scale, representative survey of practicing life scientists regarding their views on biosecurity and how they would allocate responsibility for biosecurity activities such as education, training, and oversight among individuals and institutions. The need for a baseline survey to address these questions at this time was underscored

[47]See, for example, Corneliussen (2005), McLeish and Nightingale (2005), Fischer (2006), Lentzos (2006), Sutton (2007, 2009), and U.S. Government (2008).

[48]"A significant difference between the U.S. seminars and those elsewhere was the greater knowledge about biosecurity discussions in general, potential misuse policy initiatives, and related BW issues. This is perhaps not surprising because of the heightened attention to potential misuse issues in the US and because nearly all the organizations visited had a direct stake in that they were receiving biodefence research funding" (Rappert et al. 2006).

by the release in 2007 of the NSABB's draft oversight framework for all federally funded life sciences research and its plans for education and outreach activities (NSABB 2007).

THE AAAS-NRC PROJECT

Building on the results of their 2005 planning meeting, the AAAS Center for Science, Technology, and Security Policy (CSTSP)[49] and two program units within the National Research Council[50] developed a plan to survey a sample of the AAAS membership in the life sciences about their knowledge of dual use issues and attitudes toward their responsibilities to help mitigate the risks of misuse. In addition to providing essential baseline data, it was hoped that the results of the survey would generate more attention to the continuing challenges of dual use issues and additional debate among life scientists about their personal and professional responsibilities. Concerns about potential misuse of dual use research relate to questions of whether advanced research could facilitate biological weapons development by states and nonstate actors. The introductory material in this chapter has covered both issues. The survey, however, focused on bioterrorism by nonstate actors because that is the concern that has driven so many of the policy discussions of dual use research in the United States since September 11.

The survey project occurred in three phases:

Phase 1:

- Define the issues to be addressed and identify the critical populations to be surveyed by a questionnaire that will assess the attitudes and values of a statistically valid sample of life scientists relevant to the design of biosecurity education programs; and

[49]CSTSP is a nonpartisan, nonadvocacy organization supported by the John D. and Catherine T. MacArthur Foundation's Science, Technology and Security Initiative. CSTSP serves as a two-way portal between scientists studying matters related to national and international security and the Washington policy community, especially the Congress, the executive branch agencies, and Washington-based nongovernmental organizations. The center's goal is to bring to the Washington policy community objective, high-quality, scientific and technical information related to security issues. For additional information see http://cstsp.aaas.org.

[50]NRC is part of the National Academies, which also include the National Academy of Sciences, the National Academy of Engineering, and the Institute of Medicine. Created in 1916, NRC has become the principal operating agency of both the National Academy of Sciences and the National Academy of Engineering in providing services to the government, the public, and the scientific and engineering communities. The program units involved in this project were the Board on Life Sciences and the Program on Development, Security, and Cooperation.

- Devise a preliminary questionnaire to be used in the survey.

Phase 2:

- Work with a professional firm to conduct focus groups to explore the attitudes and values of selected groups of scientists and also provide a pretest of the questionnaire;
- Implement the survey in partnership with AAAS; and
- Review the results of the survey and prepare a report of the initial findings.

Phase 3:

- Based on the analysis of the findings of the survey, recommend approaches for engaging members of the life sciences community on biosecurity issues.

NRC appointed a committee to provide oversight for the Academies' participation in the project. The biographies of the committee members can be found in Appendix A. Kavita Marfatia Berger, Project Director at the AAAS CSTSP, served as a consultant to the committee and actively participated in the development and implementation of the survey and preparation of the final report. Because funding came in increments to support particular phases of the project, however, the committee was not formally appointed until after the focus groups were completed and decisions had been made about the survey instrument, the survey design and whom to survey without input from the committee. The committee oversaw the analysis of the survey results and the preparation of the final report. To ensure the privacy of the AAAS members who were sampled, while also permitting follow-ups to encourage a higher response rate, the AAAS Office of Member Services administered the questionnaire. The NRC assumed responsibility for the review of the final report. This was a genuine partnership between the two organizations.

This chapter has provided an introduction to the issues surrounding dual use research and the rationale for undertaking the survey. The next chapter describes the survey methodology and provides details about the sample and the demographic characteristics of the respondents. Chapter 3 describes the results of the survey and Chapter 4 provides the committee's conclusions and recommendations.

2

Approach

RESEARCH OBJECTIVES

To begin to assess the attitudes of members of the life sciences community toward dual use research issues and to learn what actions life scientists in the United States would support to address them, the National Research Council (NRC) and the American Association for the Advancement of Science (AAAS) conducted a survey of a segment of the U.S. life sciences community to assess *awareness* of the dual use dilemma—including perceptions of the risk of bioterrorism, *attitudes* about responsibilities to help reduce the risks that their research could be misused, and *actions* being taken by some life scientists in response to the dual use dilemma. Specifically the survey was designed to examine:

- Views about the likelihood of bioterrorism and the potential role of dual use research in facilitating it;
- Views about the need for different types of responses, including regulation, institutional policies, or changes in individual conduct to reduce the threat of misuse of research;
- Opinions about who should play a role in education or regulation; for example, should responsibility for minimizing potential hazards from dual use research rest with scientists, professional societies, journals, research institutions, the government, or some combination;
- Information about actions that life scientists have taken in response to concerns about dual use research; and
- Whether different categories of life scientists, such as those in dif-

ferent employment sectors (e.g., academia or government or industry) have different views and opinions on these topics or have taken different actions.

STUDY DESIGN

To meet the research objectives, a questionnaire was developed and a cross-sectional Web survey was conducted. This section describes the development of the questionnaire, survey pretest, the target population and sample frame,[1] the survey mode and design, sampling issues, and implementation of the survey.

Developing the Survey Questionnaire

NRC staff developed a preliminary draft of the questionnaire. Questions were solicited from selected members of the National Academy of Sciences and NRC staff with expertise in biosecurity, staff from the National Science Advisory Board for Biosecurity (NSABB), AAAS staff, and other experts identified by their work in this area (e.g., dual use research, regulation, codes of conduct) as well as drawn from other surveys on similar topics. Once an initial set of questions was collected, it was circulated for these experts' further comments. The draft questionnaire was initially an attempt to "cast a wide net" by including as many questions as possible, leaving survey length, question order, the appropriate mix of opinion and fact-based questions, and precise wording for later in the questionnaire development. The initial questionnaire contained more than 60 questions that were identified as important; the final questionnaire contained 35 questions.

From November 2006 through March 2007, the project staff made an effort to tap the expertise of potential survey respondents as well as to bring together stakeholders to discuss issues related to biosecurity to further refine the questionnaire. In addition to continued discussion among staff, three focus group discussions were held involving junior and senior life scientists in the biological, agricultural, and medical disciplines from academia, government, and industry. The focus groups were designed to address the following:

1. Who should be responsible for (and what parts of) dual use research, including the responsibilities of scientists, journal editorial boards, and the government;

[1] A sample frame consists of a list from which individuals from the survey population can be selected.

2. A consensus on key terms and definitions, such as "dual use";

3. A sense of the range of possible answers for particular questions, such as "At what point should oversight of scientific research begin?"

Greenberg Quinlan Rosner (GQR) Research, Inc. designed and conducted the focus groups of life scientists. Two focus groups were conducted on February 7, 2007, in Bethesda, Maryland, and another on February 26, 2007, in San Francisco, California. Working with NRC and AAAS staff, GQR developed an interview script for the three focus groups, each of which had between 8 and 10 participants. The script was based on the original questionnaire.

The NRC staff identified and recruited participants to the focus groups. Participants from the first focus group included scientists from the Navy Medical Research Center, George Mason University, the National Institutes of Health (NIH), the Center for Biosecurity of the University of Pittsburgh Medical Center, the Federal Bureau of Investigation, Georgetown University, and the University of Maryland. Participants from the second group included scientists from Gryphon Scientific, Functional Genetics, the Institute for Genomic Research, NIH, the J. Craig Venter Institute, the Department of Defense, Arizona State University, and MedImmune, Inc. Participants in the final group included scientists from Lawrence Berkeley National Laboratory, Lawrence Livermore National Laboratory, Stanford University, University of California at San Francisco, University of California at Davis, and Celera Diagnostics, Inc.

The focus group discussions provided information to assist with the survey project as well as planned, future projects on biosecurity. GQR provided a summary of the group discussions (see Appendix B); their key findings included:

- The range of knowledge regarding dual use research varied widely: Several of the scientists interviewed were familiar with dual use research; some were deeply aware; but for many others, knowledge was superficial.
- There was no discernible consensus about how concerned the scientific community was about the possible misapplication of dual use research.
- This lack of consensus is rooted in views about the role science plays in society and the scientists' tolerance for risk.
- Finally, many of the life scientists interviewed, including those who felt that misuse was a pressing concern, were reluctant to sacrifice core scientific values such as transparency, open flow of information, and a desire to cure diseases in exchange for added security.

Following the focus group meetings, AAAS and NRC convened an additional small group of practicing scientists to discuss approaches for inviting principal investigators to participate in the survey. The participants included faculty from Georgetown University, George Washington University, NIH, and the University of Maryland. All the discussants at this meeting felt that the response rate would likely be low, and the suggestion was made to try to increase the response rate by sending a preinvitation announcement prior to the invitation letter. The preinvitation announcement should inform the participants about the purpose of the survey and how much time the survey would take to complete. Additionally, the Web site should inform participants of their progress in completing the survey. These recommendations were followed when distributing the survey.

After the three focus group meetings, a final questionnaire was developed and pretested, by sending it to a few dozen biosecurity experts for their comment in July 2007. These individuals tended to be more aware of dual use issues than the average respondents. Several of these experts returned comments about the survey, which were used to refine the questionnaire. The final questionnaire can be found in Appendix C.

Target Population

The target population of this study would be U.S. life scientists, particularly

- Individuals with an advanced degree in the biological sciences, the health sciences, and the agricultural sciences.
  —An advanced degree includes master's degrees, doctoral degrees, and other professional degrees (e.g., M.D. or D.V.M.). Some individuals may have joint degrees, such as a Ph.D. and an M.D.
  —"Life sciences" include the agricultural sciences and natural resources, biological and biomedical sciences, and health sciences. (An example of relevant scientific fields is the list from which recent doctorates select in the annual Survey of Earned Doctorates.[2]) The life sciences are quite interdisciplinary—a trend that is growing. For the purposes of this survey scientists based primarily in such fields as engineering, physics, or computer sciences, who might be engaged in life science research, for example via bioengineering, were excluded.[3]

---

[2]2008 Survey of Earned Doctorates Questionnaire, available at http://www.norc.org/nr/rdonlyres/7fd70cb6-6371-47a6-87fb-eaaada1734b6/0/sed07_08.pdf.

[3]This was a practical decision based on the availability of data to create a sample frame. The excluded scientists are less well represented among AAAS members.

- Life scientists working in the United States.[4]
- Life scientists working across academia, industry, and government.

However, there is no single complete list of members for this population available, and even the exact size of this population is unclear. One approach to identifying this population would be to identify the people who have received an advanced degree in life science disciplines and who are currently engaged in life sciences research in some capacity. The closest such definition of life scientists is offered by the National Science Foundation, which in 2003 found 145,760 employed scientists with doctorates in the biological, agricultural, and environmental life sciences (NSF 2006:6).[5] This count provides only an approximation of the ideal target population for this report. It does not include scientists with terminal master's degrees or scientists with foreign doctorates working in the United States.

Given the difficulties in identifying the target population, the project staff decided to find an alternative survey population that might approximate the desired characteristics of the population of U.S. life scientists, such as members of a professional scientific society or grantees from a federal agency. The professional societies considered as surrogates for the desired target population included the Federation of American Societies for Experimental Biology and its 21 member societies, the American Society for Microbiology, or Sigma Xi. Lists of grantees from government agencies such as the National Science Foundation or NIH and federal scientists at agencies such as the U.S. Department of Agriculture or the Department of Homeland Security were also considered.

After reviewing alternative survey populations, the life science members of AAAS were selected as the surrogate "target" population. AAAS is the largest general scientific society in the world; its more than 130,000 individual members include nearly 65,000 members who report that they are life scientists from one of a number of scientific fields. Since the membership is largely American (about 84 percent) and primarily composed of postgraduate scientists, there was ample opportunity to survey the attitudes of a considerable body of American life scientists in several different but related scientific fields. *Having said this, it is important to note that one cannot infer from the views of AAAS life science members to the broader life*

[4]Although issues regarding scientists' views about biosecurity affect scientists in many countries, this project in part reacts to efforts in the United States to educate scientists and to potentially set government regulations or guidelines for scientists working in the United States.

[5]Focusing instead on occupations, the Bureau of Labor Statistics calculated that for 2006 there were 245,000 employed life scientists in the United States, though this is a broader definition that, for example, is not limited to those with doctorates (BLS 2008:13–14).

science community in the United States. This severely constrains the conclusions that can be drawn from this survey.

Survey Mode and Sample Frame

Recognizing that a census of life scientists was unrealistic, given the size of the target population, the project turned to a one-time, cross-sectional survey. This methodology was chosen for procedural reasons as well as the project's desire to obtain a baseline of quantitative data on scientists' attitudes toward biosecurity and dual use research issues. Procedurally, surveys allow for standardized measurement across all respondents. As long as there is a sufficiently high response rate, a probability sample can yield unbiased estimates of population means, proportions, and totals, and allow computation of data precision as a measure of data quality without assumptions about the population distributions. A one-time survey conducted over a relatively short time period minimizes the chances that some external event will influence the respondents' answers midway through the data collection processes.[6] This type of survey also allows collection of data on the respondents, such as demographic factors, which can be related to their attitudes.[7]

Having decided on a cross-sectional survey, the project staff had to choose a survey mode from among four basic possibilities: face to face

[6]To give one example of a possible external influence, in February 2008, after completion of the survey, ricin was found in a hotel room in Las Vegas, Nevada. This incident was reported by major newspapers. In another example from August 2008, the revelations concerning scientist Bruce Ivins and the 2001 anthrax attacks received significant press coverage. In a longer data collection mode, some scientists might answer before an event such as this whereas some answer afterward.

[7]There is sparse theoretical guidance about what factors might explain differences among scientists' views or why particular factors would be relevant. Two determinants for including such variables in a survey are: (1) Is there a hypothetical reason to assume that there might be a significantly different response among subgroups and (2) what is the benefit of asking a demographic question in light of concerns about anonymity? Current surveys of scientists focusing on other issues (e.g., ethics, communication, and conflicts of interest) have employed a wide range of demographic characteristics, with mixed results. Korenman et al. (1998:45) found that respondents' "sex, age, academic rank, and scientific field were not associated with a meaningful difference in malfeasance ratings." The study of scientific communication by the Royal Society, the Research Councils UK, and The Wellcome Trust (2006) sampled faculty disaggregated by discipline and rank and included questions on age, sex, and race and ethnicity. That survey found that age or rank and discipline were relevant explanatory predictors of public scientific communication. Martinson et al. (2005) found that seniority of scientists—early versus mid-career—was a relevant factor in ethical behavior. Martinson et al. (2006) examined the effect of sex, marital status, and age, among other factors in their survey of scientists' perceptions of organizational justice. We felt that three characteristics (age, employment, and field) would be the most useful.

interviews, telephone surveys, mail surveys, and Internet surveys (either e-mail or via the Web).[8] Each method has known advantages and disadvantages (Fowler 2001; Owens 2005). Selection of an appropriate survey mode is based on a variety of factors, including the nature of the population, sample size, types of questions, research topic, projected response rates, cost, and time. For this study, primarily for reasons of cost and time, it was decided to conduct a Web-based survey.[9]

The decision to use a Web-based survey had a major impact on the creation of a sample frame (the list of individuals drawn from the survey population). Ideally, the sample frame would be the same as the survey population, but they are seldom identical. Web surveys require a frame that has e-mail addresses available for all elements. But, only slightly more than one-third of the AAAS members in the life sciences (24,194) had provided their e-mail addresses to the association. The result was that the sample frame was drawn only from AAAS members in the life sciences with available e-mail addresses. Thus, a potential bias was introduced into the study because there was a discrepancy between the sample frame of AAAS life science members with e-mail addresses and the survey population of all AAAS life sciences members. This discrepancy adds to the uncertainties about whether the results of the survey can be generalized and significantly limits any inferences that can be made about the data collected.

Basically the problem is that the responses of the 40,000 members who did not provide current e-mail contact information might differ from the approximately 24,000 who did. There could be a real difference between those with and those without known e-mail addresses. Table 2-1 illustrates the differences between the survey population of AAAS life scientists and

[8]It is also possible to conduct mixed-mode surveys, using more than one of these types, for example, by giving respondents the choice of completing a questionnaire on hard copy (mailed) or on the Web. Additionally, follow-up requests to nonrespondents can utilize a different mode. For example, a researcher could call nonresponders to a Web-based survey and conduct telephone interviews.

[9]In general, advantages of using Web-based surveys include database collection of responses (minimizes human error), no interviewer bias, and rapid data collection. Further, Web surveys allow programmed skip patterns, which reduce the chance that a respondent may answer the wrong questions, something that may arise in paper-and-pencil self-administered surveys. The major disadvantages of a Web-based survey are incomplete population coverage, low response rates, risk of technical problems, and personal security concerns. Web-based surveys thus face technological impediments (e.g., Does everyone in the sample have access to the Internet? Are they using the same Web browser?) and security issues (e.g., e-mails to potential respondents being mistaken for spam and deleted unopened or security settings preventing the use of tracking cookies sometimes used in completing surveys). Practicing life scientists are likely to have access to the Internet, limiting the technological issues, and to understand the ability of competent survey professionals to reduce the risk of disclosure through careful survey management.

TABLE 2-1 Comparison of AAAS Life Scientists in the Survey Population, the Sampling Frame, and Those Sampled, by Scientific Field

	Survey Population (AAAS members in the life sciences)		Sample Frame (AAAS members in the life sciences with known e-mail)		Sample Surveyed (AAAS members in the life sciences with known e-mail sent questionnaire)	
Scientific Field	Number	%	Number	%	Number	%
Medicine	9,303	14.4	3,416	14.1	1,418	14.2
Biochemistry	7,617	11.8	2,835	11.7	1,177	11.8
Molecular biology	6,381	9.8	2,416	10.0	1,003	10
Neuroscience	5,798	8.9	2,086	8.6	865	8.7
Cell biology	4,761	7.3	1,794	7.4	745	7.5
Biotechnology	3,962	6.1	1,745	7.2	724	7.2
Genetics	3,687	5.7	1,485	6.1	616	6.2
Microbiology	3,695	5.7	1,454	6.0	603	6.0
Immunology	3,337	5.2	1,264	5.2	525	5.3
Other life science	4,090	6.3	1,257	5.2	522	5.2
Ecology	2,911	4.5	1,043	4.3	431	4.3
Endocrinology/ Physiology	2,502	3.9	972	4.0	403	4.0
Pharmacology	2,200	3.4	812	3.4	338	3.4
Agricultural science	1,684	2.6	632	2.6	262	2.6
Zoology	1,117	1.7	380	1.6	158	1.6
Botany	882	1.4	326	1.3	114	1.1
Marine biology	860	1.3	277	1.1	96	1.0
Total	64,787	100.0	24,194	100.0	10,000	100.0

SOURCE: AAAS Member Services; calculations by staff.

the sample frame based upon scientific field. Even though the percentages in each field among the survey population of 64,787 closely match those in the sample frame of 24,194, extrapolation beyond the sampling frame for other measures would still be tenuous. Unfortunately, there were no other fields upon which the similarities between the survey population and the sample frame could be determined. Thus, there is no way to tell whether the sample frame is really representative of the survey population or the population of all U.S. life scientists.

Ten thousand AAAS members from the sample frame of 24,194 AAAS members with known e-mail addresses were selected systematically with

a random start.[10] The distribution of individual life scientists in the sample frame and in the sample are presented by discipline in Table 2-1. The important point illustrated in the table is that the percentage of life scientists in an individual field is similar across all three groups; for example, it is 14 percent for AAAS life scientists in medicine, for AAAS life scientists in medicine with known e-mail addresses, and for AAAS life scientists in medicine with known e-mail addresses who were sent the survey.

Typically, researchers have some information (such as scientific field) about individuals in the sample who did not respond, but that was not the case here. To protect confidentiality and anonymity, AAAS Member Services—who fielded the survey—delinked the names of their members in the sample frame from the outcome. After the survey's completion, it was not possible to recover information about who the respondents were. This delinking reduced the committee's ability to investigate potential nonresponse bias, again constraining the ability to generalize from the respondents to even the sample frame. *As a result of the issues relating to whether the surveyed sample was biased, which is compounded by the low response rate that will be discussed later, the committee adopted a conservative approach of reporting the results based upon the raw data provided by the respondents rather than inferring to the sample frame or survey population.* Consequently, only inferences for further investigation can be made because of the limitations of the survey design and response.

Implementation and Survey Response

The survey was fielded from August 8 to October 12, 2007. Ten thousand AAAS members were contacted via e-mail by AAAS Members Services.[11] All individuals included in the survey were sent a preliminary announcement to inform them of the NRC/AAAS survey on dual use research and were informed that the survey link would be sent after 2 days. On August 8, the invitation letter containing the survey link was sent to all 10,000 selected AAAS members. Six follow-up e-mails were sent, one approximately every 2 weeks.

The survey Web site was visited by 2,713 among the 10,000 sampled

[10]The 10,000 scientists were selected by (1) numbering the list of 24,194 scientists with e-mail addresses; (2) selecting a random number between 1 and 24; (3) adding 2.4 to that number until we had a list of 10,000 numbers; and (4) selecting individuals that corresponded to those numbers. For example, if the random number was 7, then the beginning of the sequence would be: 7, 9.4, 11.8, 14.2 . . . and the 7th, 9th, 12th, and 14th persons would be selected. Unfortunately, the number should have been 2.4194. Truncating the decimal point meant that the last 170 people on the list could not have been contacted.

[11]This may have included individuals from the focus groups or those contacted during development of the survey and pretesting.

TABLE 2-2 Response Rate Calculator

Type of Respondent	Number	Percentage
Responded		
Completely	1,570	15.7
Partially	384	3.8
Did not respond	7,762	77.6
Not eligible (nonworking e-mail)	284	2.8
TOTAL	10,000	100.0

SOURCE: Calculations by staff.

scientists, a 27 percent contact rate. A total of 1,954 completed part of the survey and 1,570 completed the entire survey, as shown in Table 2-2.

This leads to a response rate of about 16 percent for completed surveys, 20 percent if partial responses are included.[12]

POTENTIAL SOURCES OF ERROR

There are a number of potential sources of error in any survey, including sampling error, measurement error, coverage error, nonresponse error at the unit (individuals not responding to the questionnaire) and item level (individuals not answering some of the questions), and postsurvey error (Weisberg 2005). Unit nonresponse, item nonresponse, and measurement error are discussed subsequently.

Life Scientists Did Not Complete the Survey (Unit Nonresponse Error)

Many individuals who are asked to complete a survey fail to do so. In general, survey response rates have been dropping over time, and in response, a number of studies have examined different strategies to maximize response rates, including design issues, timing issues, and the use

[12]Surveys include four types of persons: those who returned the questionnaire (respondents); those who were contacted, but did not respond for some reason (e.g., refusals or people who were unavailable when the survey was fielded)—these are nonrespondents; people whose eligibility cannot be determined (e.g., never logged on to an Internet survey or bad e-mail addresses); and people who were ineligible (e.g., duplicate listing) (AAPOR 2006). There are multiple ways to define a survey's response rate. Generally, the response rate is defined as the number of individuals for which an attempt was made to collect data, who are members of AAAS, and who responded to the survey divided by the number of individuals who were eligible to be sampled multiplied by 100; this rate is expressed as a percentage. The numerator may include only complete responses or both complete and partial responses. The denominator includes all responders, nonresponders, and unknowns. See AAPOR (2006) for a more thorough discussion of different definitions of response rates and other rates of relevance.

of incentives. Web-based surveys do not always have high response rates. Cook et al. (2000) published a meta-analysis of 68 Web or Internet surveys that had a mean response rate of 39.6 percent (with a standard deviation of 19.6 percent).[13] A more recent report by the RAND Corporation, which assessed several surveys of different modes, found that response rates for Web surveys ranged between 7 and 44 percent (Schonlau et al. 2002). According to conversations with AAAS Member Services, response rates of 5 to 10 percent to large AAAS Web-based surveys with more than 30 questions are common.

The response rate for this survey was thus probably higher than average for AAAS surveys, but certainly on the mid to lower end for Web-based surveys in general. There are several potential explanations. One is survey sponsorship. Respondents who either did not read the e-mail message sent by AAAS or read the message but did not follow the link to the survey might have mistaken the survey for a request from AAAS (e.g., an offer to renew membership). Second, some respondents may have decided the survey topic was not sufficiently interesting and they deleted or ignored the e-mail. Lack of interest may have been compounded because individuals may have already received similar requests to participate in surveys. At the final focus group, one interviewee noted that he had already received more than a dozen requests to fill out a survey in the first 6 months of the year and he was less inclined to fill out surveys as a result. Finally, the length of the questionnaire may have dissuaded respondents.

Is a low response rate a problem for generalizing the results of this survey? Thresholds for acceptable response rates are not definitive. Lower response rates increase the risk that there will be nonresponse bias, arising when individuals who do not respond to the survey have different answers from those who did answer. The differences could lead to over- or underestimates in results based only on respondents, and those differences can vary by question. Low response rates, however, in and of themselves do not necessarily mean that bias in survey estimates exists (Curtin et al. 2000; Keeter et al. 2000; Merkle and Edelman 2002: Groves 2006).

Where bias is presumed to exist, it can be conceptualized in several ways. Groves (2006) suggests that one way to think about nonresponse bias is to consider the correlation between an individual's propensity to respond with the attributes being measured via the survey. He identifies five models relevant to the covariance between respondents' probability of responding to a survey and their responses to it. One in particular seems especially relevant to this project: the "common cause model,"

[13]Note that this analysis includes both e-mail and Web surveys.

which "generates a covariance between the two attributes because of a common cause of both of them" (Groves 2006:650). Applied to this particular survey, this model may suggest that some individuals in the sample were (1) more likely to respond to the survey and (2) this interest influenced some survey variables (e.g., whether they supported particular regulatory actions). Hypothetically, it could be the case that individuals who worked with select agents and thus were more familiar with regulations would (1) be more likely to respond (because the issue itself or the potential burden of additional regulation was more important to them), and (2) tend to have similar views (e.g., hypothetically they might be more accepting of additional regulation). If this hypothesis were correct, then for this question the estimated percentage of scientists accepting regulation would be too high, because the sample included a disproportionate share of life scientists working with select agents and willing to accept oversight measures.

There are a number of factors that might have made certain scientists more or less likely to take the NRC/AAAS survey (e.g., field, kind of research, general attitudes about regulation). Unfortunately, there is little evidence available to the committee to assess nonresponse bias in this survey. There are several ways nonresponse bias could have been assessed, but, as described above, because of confidentiality and anonymity concerns, the necessary frame data were not available from AAAS. Another approach to assess nonresponse bias would have been to compare the results of the committee's survey with results from similar surveys of life scientists. As noted in Chapter 1, similar surveys of a broad sample of life scientists do not exist, although in Chapter 3 we discuss how the results from this survey compare to some work targeting the public and policy makers, and a few studies of groups of scientists engaged in biodefense research. A third approach would be to conduct a study of nonrespondents. One could, for example, telephone nonrespondents or mail them the survey instrument, if they were known and their contact information was available. Cost constraints and the difficulty of identifying nonrespondents meant that this type of follow-up was not possible.

Readers should thus be aware that the size of the nonresponse bias is unknown in this study. Caution should be applied in interpreting the findings presented in the next chapter. The committee sees this survey as a first effort to gauge scientists' views on this important topic. It should be viewed as generating hypotheses rather than providing conclusive results.

Life Scientists Did Not Complete the Entire Survey (Item Nonresponse Error)

The survey was quite long and many respondents started but did not finish the survey. A total of 1,950 respondents answered the first question: "Have you ever conducted research or managed others' research in the life sciences?" For questions 2–12 (the first third of the questionnaire), the number of responses ranges from about 1,700 to 1,800 responses. For questions 13–23, the number of responses drops into the 1,600s. For the remaining questions (24–35), the number drops again, with one exception, into the 1,500s. Response rates for individual questions are presented in Appendix D (Table D-1). The fact that many respondents did not answer every question also adds uncertainty to the results.

Measurement Error

As Dillman (1991:228) explains: "Measurement error refers to the discrepancy between underlying, unobserved variables (whether opinions, behaviors, or attributes) and the observed survey responses. Whereas the three preceding types of errors (sampling, noncoverage, and nonresponse) stem from nonobservation, measurement error results from the process of observation." That is, measurement error can result from the characteristics of respondents, for example, an inability to recall events or provide accurate information. "Measurement error may also result from characteristics of the question (e.g., a question phrased so that it cannot be answered correctly) or of the questionnaire (e.g., the order in which questions are presented)" Dillman (1991:228).

In the NRC/AAAS survey, measurement error may exist for two important questions. Question 5, which asked respondents if their research fit into one of seven categories of experiments (defined by the NSABB), was prefaced with the definition of these categories. This preface was long and technical, and some respondents may not have read it. Question 6 asked respondents whether they had ever worked with select agents. As will be discussed further in the next chapter, the number of respondents who said "Yes" is very difficult to explain unless the question was not properly understood or there was bias in who chose to answer the question.

DATA ANALYSIS

Analytical Approach

This report uses two approaches to reporting the results of the survey. First, descriptive statistics are used to report frequencies of answers

to particular questions. Later in this chapter, for instance, we present the percentage of scientists holding different types of terminal degrees (e.g., masters, Ph.D., M.D.). Measures of central tendency, such as the mean or median, are also presented. Less frequently, we present cross-tabulations between selected variables to further disaggregate life scientists into smaller groups. For example, where are scientists who work with select agents employed—in government, academia, industry, or somewhere else?

The second type of analysis used in this report consists of correlations, detailing the relationship between pairs of variables. Three correlation statistics are employed: the Pearson correlation coefficient (r), which is used for two interval variables;[14] Spearman's rho (ρ), which is used when two variables are interval (e.g., Likert scale variables such as 1 to 5 where 1 is strongly disagree and 5 is strongly agree),[15] and phi (ϕ), which is used when the two variables are binary (e.g., Yes/No).

CHARACTERISTICS OF RESPONDENTS

The results of the survey are presented in Chapter 3, but this section describes the data collected by the NRC/AAAS survey on background characteristics of respondents that were hypothesized to be relevant to help explain responses to other survey questions. Almost all of the respondents had conducted or managed life sciences research, were employed, had a postgraduate degree, and were U.S. citizens. A majority of scientists were academics and most were mid-career.

Since one goal for the survey was to discern and describe attitudes and opinions of scientists who had some experience with research, the first survey question was: "Have you ever conducted research or managed others' research in the life sciences?" A related question was: "Are you currently conducting or managing research in the life sciences?" As Table 2-3 illustrates, almost all of those who answered the first question answered positively; and a majority of those who said "Yes" to the first question were also currently involved in research at the time of the survey.

Table 2-3 also shows that 95 percent of respondents (1,861 of 1,950) who answered the question had conducted or managed research at some point in their careers. Among those respondents who answered affirmatively, 80 percent (1,407 of 1,758) of those who also answered the second question were currently conducting or managing research. Additionally,

---

[14]Pearson r assumes a normal distribution.

[15]Spearman ρ is a nonparametric measure of correlation that does not assume a particular frequency distribution.

TABLE 2-3 Respondents' Current Role in Scientific Research

Respondent Has Ever Conducted or Managed Research	Respondent Is Currently Conducting or Managing Research			
	Yes	No	Did Not Answer	Total
Yes	1,407	351	103	1,861
No	0	85	4	89
Did not answer	0	0	4	4
Total	1,407	436	111	1,954

NOTE: Fourteen answers were reassigned; 14 respondents said they had never conducted or managed research, but then said they were doing so currently. These were changed from "No" to "Yes" for the question: Have you ever conducted or managed research.
SOURCE: NRC-AAAS Survey; data tabulations by staff.

TABLE 2-4 Employment Status of Respondents

Status	Frequency	Percentage
Employed	1,464	92
Retired	92	6
Unemployed	16	1
Other	14	1
Respondents	1,586	100
Did not answer	368	—
Total	1,954	—

SOURCE: NRC-AAAS Survey; data tabulations by staff.

the majority of life scientists who responded to the survey have conducted research since 2001, which is the period during which the greatest concerns have been expressed about the potential for misuse of the life sciences to aid bioterrorism.

Among those life scientists who answered the question on their employment status, most (92 percent) were employed when the survey was fielded, as shown in Table 2-4.

As noted in Table 2-5, most respondents who selected an employment sector were employed in academia (71 percent), followed by those employed in the commercial sector (16 percent) and government employees (9 percent).[16] There also were some employed in other unspecified categories, such as contractors at government labs, employees of

[16]This includes both federal and nonfederal government employees.

TABLE 2-5 Employment Sector of Respondents

Employment Sector	Frequency	Percentage
Academic	1,023	71
Industry	223	16
Government	125	9
Other	72	5
Respondents	1,443	100
Did not answer	511	—
Total	1,954	—

SOURCE: NRC/AAAS Survey; data tabulations by staff.

health care facilities, and employees of nongovernmental organizations (5 percent).[17]

Table 2-6 shows that the life scientists who responded to the survey represented a wide range of subfields. Most subdisciplines represented less than 10 percent of the total population. Fields with the most respondents were biochemistry, medicine, microbiology, and molecular biology.

The life scientists who responded to the survey overwhelmingly had postgraduate degrees (79 percent had a Ph.D., while 6 percent had a joint Ph.D. and another professional degree) as noted in Table 2-7. Nine percent had another professional degree such as an M.D., D.V.M., or J.D.

Figure 2-1 shows the wide range of career experience of the life scientists who responded to the survey—from very junior scientists to those who had retired. For the 1,586 respondents who answered the question about highest educational degree awarded, the mean years since highest degree awarded was 23 and the median years since highest degree awarded was 24, indicating that the life scientists included in this study were generally mid-career or higher.

Finally, as noted in Table 2-8, just under 97 percent of the respondents who answered the question on citizenship indicated that they were U.S. citizens or permanent residents, while only about 4 percent answered that they were non-U.S. citizens. This result was expected since the survey

[17]The AAAS Member Services uses a different categorization scheme. According to the AAAS system, the employment sectors for the 24,194 members included in this study were 54.2 percent University/College, 15.2 percent Industry/Business, 10 percent Healthcare, 7.2 percent Government, 6.6 percent Nonprofit organization, 5.9 percent Other (e.g., contractors), 0.5 percent unknown, and 0.1 percent each for Hospital, Student, and Retired. However, because of confidentiality concerns, it is not known what the AAAS classification scheme would have yielded for the 10,000 in the sample. It is interesting, but unclear, as to why a considerably higher proportion of academic scientists responded than one might have expected from the AAAS classification.

TABLE 2-6 Scientific Field of Respondents

Field	Percentage[a]
Biochemistry	12
Medicine	10
Microbiology	10
Molecular biology	10
Neuroscience	9
Cell biology	8
Other life sciences	8
Genetics	7
Biotechnology	6
Immunology	5
Ecology	4
Agricultural science	3
Pharmacology	3
Endocrinology/physiology	2
Marine biology	2
Zoology	2
Botany	1
Respondents	1,586
Did not answer	368
Total	1,954

[a]Does not equal 100 percent due to rounding.
SOURCE: NRC/AAAS Survey; data tabulations by staff.

TABLE 2-7 Highest Awarded Degree of Respondents

Highest Awarded Degree	Frequency	Percentage
Doctorate or equivalent (e.g., Ph.D., D.Sc., Ed.D.)	1,252	79
Other professional degree (e.g., J.D., L.L.B., M.D., D.D.S., D.V.M.)	143	9
Joint doctorate and professional degree (e.g., Ph.D. and M.D.)	89	6
Master's degree or equivalent (e.g., M.S., M.A., M.B.A.)	64	4
Bachelor's degree or equivalent (e.g., B.S., B.A., A.B.)	34	2
Other	4	0
Respondents	1,586	100
Did not answer	368	—
Total	1,954	—

SOURCE: NRC/AAAS Survey; data tabulations by staff.

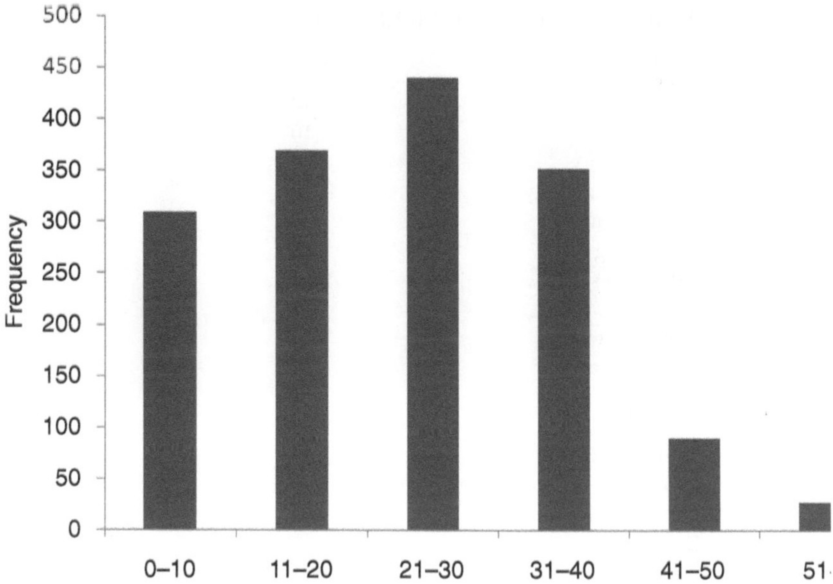

FIGURE 2-1 Years since highest degree was awarded.
NOTE: Based on 1,586 responses.
SOURCE: NRC/AAAS Survey; data tabulations by staff.

TABLE 2-8 Citizenship of Respondents

Citizenship Status	Frequency	Percentage[a]
U.S. citizen since birth or naturalized	1,440	91
Non-U.S. citizen, with a Permanent U.S. Resident Visa (Green Card)	88	6
Non-U.S. citizen, with a Temporary U.S. Resident Visa	58	4
Respondents	1,586	101
Did not answer	368	—
Total	1,954	—

[a]Does not equal 100 percent due to rounding.
SOURCE: NRC/AAAS Survey; data tabulations by staff.

focused on scientists working in the United States by selecting the sample from those AAAS members with U.S. e-mail addresses.

CONCLUDING REMARKS

The survey was undertaken to assess the attitudes and opinions of a large group of life scientists because such information is not currently

known and could help in the development and implementation of policies and activities to address dual use research. However, given the difficulty of finding a representative survey group from which to sample and the low survey response rate, these ambitious goals could not be fully achieved. Only 16 percent of scientists completed the survey and another 4 percent responded in part. Nevertheless, the committee believes that the survey responses (including respondents' comments) do provide interesting indications of attitudes and evidence of actions that merit further investigation. Moreover, the committee feels that the methodological difficulties encountered in this project serve as valuable lessons for future surveys on this as well as other topics of interest with the scientific community.

[illegible] could identify the [illegible] and [illegible] of the [illegible] to address [illegible] High, however, [illegible] faculty [illegible] a representative survey [illegible] and the [illegible] survey [illegible] these [illegible] fully [illegible] rate [illegible] percent [illegible] completed the survey and another [illegible] percent responded in part. Nevertheless, the committee believes that the survey [illegible] including responses [illegible] comments) does provide interesting indications of attitudes and [illegible] of [illegible] that merit [illegible] consideration [illegible] [illegible] [illegible] [illegible] this present survey might be [illegible] [illegible] on this [illegible] other forms of [illegible] in the [illegible]

3

Results of the Survey

This chapter presents the results of a survey of life scientist members of the AAAS who could readily be reached via e-mail about a range of issues related to biosecurity: their perceptions of the risk of bioterrorism and whether aspects of scientific communication might somehow contribute to increasing those risks; whether they have made changes in how they conduct research in response to dual use concerns; and their attitudes about what policies and practices might reduce the risks of dual use issues and who should have responsibility for implementing them. Quantitative data were collected via a Web-based survey, with invitations to participate sent to a sample of 10,000 life scientists who are AAAS members with verified e-mail addresses. Of the 10,000 scientists in the sample, 1,570 (or 15.7 percent) responded with completed questionnaires. Almost all of the respondents have conducted or managed life sciences research (and three-quarters of them are currently doing so), were employed, had a postgraduate degree, and were U.S. citizens. In addition, as described in Chapter 2, a substantial majority of the scientists were academics and most were mid-career.

This chapter is divided into several sections reflecting particular results from the survey. Each section contains a brief background and a discussion of the responses and their possible implications, followed by a list summarizing the key results. The next section focuses on the types of research that the scientists who responded to the survey conduct, such as dual use or work with select agents. That is followed by a section on the views of these life scientists on bioterrorism, which includes both

their perceptions of the biosecurity risk and their views about a variety of conditions that could facilitate a biorterror attack. The questionnaire was organized around the potential responsibilities of individuals and institutions so the presentation of results in the "Responsibility for Oversight" section follows that structure. The final section on "Policy" is focused on what steps the life scientists who responded to the survey support to reduce the potential that research results could pose a threat to national security. This analysis yielded results that the committee believes can help inform educational and outreach efforts to increase awareness and enhance oversight of research that raises concerns about dual use potential.

RESPONDENTS' RESEARCH EXPERIENCE

Background

As noted in Chapter 2, about 95 percent of the respondents to the survey had conducted or managed research at some point in their careers and 80 percent of those who had were also currently conducting or managing research. This section focuses on the types of research that these scientists conduct, specifically whether they considered their research to be dual use, their research involved any of seven categories of experiments identified by the National Science Advisory Board on Biosecurity (NSABB 2007), or their research involved select agents. Each of these types of research has been identified as producing the sort of results that raise concerns about potential misuse. The terms are discussed below, including an explanation of the definitions provided to survey participants.

Asking scientists about their experiences with dual use research is complicated by the fact that the term "dual use" can have different definitions.[1] The committee was not certain if scientists were familiar with the concept of dual use research of concern as it is used by the National Science Advisory Board for Biosecurity (NSABB 2007). To help avoid confusion, the survey provided the following definition of dual use research: "In recent years, members of the scientific and security policy communities have raised concerns about the potential for misuse of knowledge, tools and techniques for purposes of bioterrorism. Such research is sometimes called 'dual use' research because, although the research is intended only for beneficial purposes, it could be misapplied." On the basis of this definition, survey participants were asked whether they considered any

[1]See Atlas and Dando (2006) for a discussion of the various meanings of the term "dual use."

of the research they were currently conducting or managing to have "dual use potential"[2]

Another way to consider whether scientists are carrying out dual use research is to focus on particular types of experiments that seem most likely to raise issues relative to the potential for misuse. This approach was recommended by the Fink committee in its National Research Council (NRC) report, *Biotechnology Research in an Age of Terrorism*:

> The Committee identified seven classes of experiments that it believes illustrate the types of endeavors or discoveries that will require review and discussion by informed members of the scientific and medical community before they are undertaken or, if carried out, before they are published in full detail. These categories represent experiments that are feasible with existing knowledge and technologies or with advances that the Committee could anticipate occurring in the near future (NRC 2004a:113).

The National Science Advisory Board for Biosecurity (NSABB), building on the Fink report, adopted a similar list of seven categories of experiments (see Box 1-2), which focus on microbial threats (NRC 2004a). Since this is the list that would form the basis for the policies recommended by the NSABB for all federally funded dual use research of concern (NSABB 2007), the committee wanted to learn how many of the respondents were conducting this type of research. Because the level of awareness of the NSABB list among the U.S. scientific community is not known, the question was prefaced with the following introduction:

> The National Science Advisory Board for Biosecurity (NSABB) has identified a subset of life sciences research that they believe may be worthwhile but may also need special review. Such research includes experiments designed to (1) enhance the harmful consequences of a biological agent or toxin; (2) disrupt immunity or the effectiveness of an immunization without clinical and/or agricultural justification; (3) confer to a biological agent or toxin, resistance to clinically and/or agriculturally useful prophylactic or therapeutic interventions against that agent or toxin, or facilitate their ability to evade detection methodologies; (4) increase the stability, transmissibility, or the ability to disseminate a biological agent or toxin; (5) alter the host range or tropism of a biological agent or toxin; (6) enhance the susceptibility of a host population; and (7) generate a novel pathogenic agent or toxin, or reconstitute an eradicated or extinct biological agent (NSABB 2006).[3]

---

[2]This question was only asked of respondents who were currently conducting or managing research.

[3]These were the categories specified by the NSABB at the time of the survey.

Finally, while dual use research of concern is not a category currently used to trigger formal federal oversight, scientists in the United States who currently work with "select agents" are subject to significant government oversight aimed at enhancing biosecurity. The Centers for Disease Control and Prevention (CDC) regulate the possession, use, and transfer of select agents and toxins that have the potential to pose a "severe threat to public health and safety," and register all laboratories and other entities in the United States involved in such activities. The Animal and Plant Health Inspection Service (APHIS) administers a comparable program for agents and toxins that could threaten plant and animal health.[4] Experience with select agents could affect one's views on dual use research and potential oversight measures.[5] The survey asked participants if they had ever worked with or managed research using select agents.[6] The questionnaire offered the following definition for select agents: "a microorganism (virus, bacterium, fungus, rickettsia) or toxin listed by the CDC and the USDA as harmful to public or agricultural health."[7]

Results and Discussion

Among those life scientists who indicated that they are currently engaged in research, 16 percent considered their research to have dual use potential. Only a few scientists who responded to the survey thought that their current research fit into one (or more) of the seven NSABB categories of experiments. One-quarter of the respondents had worked with select agents at some point in their careers, as noted in Table 3-1.

Overall, 25 percent of the respondents said that they were working with or had previously worked with select agents. The committee thought that this seemed a rather high percentage given its understanding of those currently working with select agents. To give a sense of the scope of the current programs, there are approximately 400 high-containment facilities

[4]More information on the CDC and APHIS programs can be found at http://www.cdc.gov/od/sap/.

[5]Note that as described by the NSABB, the criteria for dual use research apply to a much broader category of life sciences research, although they may incorporate certain research projects that involve select agents, depending on the nature of the particular experiments and the potential for misuse of the results and/or technology (http://oba.od.nih.gov/biosecurity/nsabb_faq.html#NSABB_FAQ016).

[6]Information about who is working with select agents is excluded from FOIA as it is seen as a potential security risk and there was concern that if the survey asked who was currently working with select agents individuals would not have answered the question out of concern that this was trying to find information that should not be revealed.

[7]This may have been interpreted as a broader definition of select agents, including those regulated agents based upon terrorism potential as well as potential harm to public or agricultural health.

TABLE 3-1 Number of Respondents Who Consider Their Research Dual Use, to Involve One of the Seven Categories of Experiments, or to Involve Select Agents

Variable	Yes	No	Don't Know	Did Not Answer	Total
Current research is dual use	215 (16%)	1,161	NA	31	1,407[a]
Current research involves one of the seven categories of experiments	82 (5%)	1,294	NA	31	1,407[a]
Have worked with select agents	458 (25%)	1,200	140	156	1,954

[a]Question only applies to those respondents who answered that they were currently conducting or managing research. Percentages are the proportion of those answering "Yes" among those who gave any answer.
SOURCE: NRC/AAAS Survey; data tabulations by staff.

currently registered with the CDC to work with select agents, of which 37 percent are nonfederal, 20 percent are federal, 8 percent are commercial, 30 percent are academic, and 5 percent are private.[8] The Federal Bureau of Investigation had cleared some 15,000 individuals to work in these facilities as of December 31, 2007.[9] According to the CDC, about 12,000 people are actively working with select agents today.[10]

Using these figures for the numbers of individuals who have been cleared to work with select agents and are estimated to be actively working with those agents today, it is reasonable to conclude that fewer than 10 percent of working life scientists in the United States could legally be working with select agents.[11] Thus, the overall number of life scientists

[8]The figures for 2007 are from the CDC Office for Terrorism Preparedness and Emergency Response; similar figures appear in the report of the Commission on the Prevention of WMD Proliferation and Terrorism (2008:25).

[9]CDC Office for Terrorism Preparedness and Emergency Response. This figure includes more than doctoral-level researchers; clearances are also required for laboratory technicians and other nonscientific personnel.

[10]Data presented at the December 10, 2008 meeting of the NSABB by Robbin Weyant, Director, Division of Select Agents and Toxins, CDC, Department of Health and Human Services.

[11]As discussed in Chapter 2, although there is no complete list of the members of the population of life scientists in the United States available, in 2003 the National Science Foundation found 145,760 employed scientists with doctorates in agricultural, biological, and environmental life sciences (NSF 2006:6).

working with select agents requiring registration must be well under the 25 percent reported by respondents to the survey.

There are several possible explanations for the high number of survey participants who reported working with select agents. First the question was worded so that individuals who had worked with agents on the select agent list prior to 2002, when registration was mandated, would be included. Moreover, it is possible that some respondents did not really know what was on the select agent list—the survey did not actually provide a full list of agents. In addition, clinical diagnostic laboratories are exempt from registration if cultures are destroyed or transferred to a registered laboratory within a 7-day period following diagnosis. Thus, clinical microbiologists and others working in laboratories may have worked with select agents but not have done research on those agents and thus not have been required to have select agent clearance. Also, those life scientists working with toxins on the select agent list below threshold concentrations do not need to be registered—many, perhaps even most, neurobiologists have worked with tetrodotoxin, botulinum toxin, and other agents in neurophysiology experiments.[12]

Finally, it is possible that the respondents represented a biased portion of the population included in the study, that is, that those individuals who currently or previously worked with select agents and performed biosecurity-related research were more likely to have answered the questions or to have responded to the survey than those who considered the survey topic remote from their area of research interest.

As noted in Table 3-2, some scientists selected each potential combination resulting from the three types of research, making for a complex relationship among the three types of research that may have influenced responses to the survey regarding acceptability of oversight and actions that may have been taken in response to dual use concerns.

The combinations above give rise to one particularly interesting result: One might have expected that anyone who answered "Yes" to working with the seven NSABB categories of experiments would also answer "Yes" to working with dual use research, but this was not the case. Dual use as it is defined in the questionnaire was clearly interpreted more broadly than the 7 categories of experiments specified by the NSABB.

[12]Under the current regulations, entities that do not at any time have more than the following aggregate amounts (in the purified form or in combinations of pure and impure forms) under the control of a principal investigator are excluded from requirements of the regulation: abrin (100 mg), botulinum neurotoxin (0.5 mg), *Clostridium perfringens* epsilon toxin (100 mg), conotoxins (100 mg), diacetoxyscirpenol (1,000 mg), ricin (100 mg), saxitoxin (100 mg), shiga-like ribosome inactivating proteins (100 mg), shigatoxin (100 mg), staphylococcal enterotoxin (5 mg), tetrodotoxin (100 mg), and T-2 (1,000 mg) (http://www.cdc.gov/od/sap/).

TABLE 3-2 Percentage of Respondents Working on Each Type of Research

Works with Dual Use Research	Works with Seven Categories of Experiments	Work with Select Agents	Percentage
No	No	No	64
No	No	Yes	20
No	Yes	No	1
No	Yes	Yes	1
Yes	No	No	5
Yes	No	Yes	5
Yes	Yes	No	2
Yes	Yes	Yes	2
Total			100
Respondents[a]			1,260
Did not answer[b]			694
Total			1,954

[a]Only includes respondents who answered "Yes" or "No" to all three questions.
[b]Includes those who did not answer and logical skips.

Interestingly, Table 3-3 shows that for the 82 respondents who reported that their research involved one of the seven categories of experiments (i.e., answered "Yes," as indicated in Table 3-1), 30 percent of those did not consider their research to be dual use. This could be explained by the fact that the seven categories are intended as a way to identify research with *dual use potential*. Or it could indicate disagreement with the Fink committee and NSABB categories as representing the full range of dual use research. Although experiments in these categories would be subject to an initial review under the draft recommendations of the NSABB, neither the NRC report nor the NSABB assumes that every experiment that met the criteria for inclusion in one of the seven categories would be considered dual use research (NRC 2004a; NSABB 2007).

TABLE 3-3 Percentage of Respondents Whose Current Research Was Dual Use and Was One of Seven Categories of Experiments

Research is Dual Use?	Research in One of the 7 Categories of Experiments?		
	Yes	No	Total
Yes	4	11	15
No	2	83	85
Total	6	94	N = 1,376

NOTE: Only includes respondents who answered both questions.
SOURCE: NRC/AAAS Survey; data tabulations by staff.

TABLE 3-4 Number and Percentage of Respondents Whose Research Involves Dual Use, Seven Categories, or Select Agents, by Employment Sector

Employment Sector	N	Percentage[a]	Respondents
A. Work involves dual use research			
Industry	38	26	147
Academia	117	13	919
Government	25	31	81
Other	10	23	44
Total	190	16	1,191
B. Research involves one or more of the seven categories of experiments			
Industry	9	6	147
Academia	53	6	919
Government	9	11	81
Other	1	2	44
Total	72	6	1,191
C. Work involves or has involved select agents			
Industry	53	26	205
Academia	260	27	959
Government	46	39	117
Other	19	29	65
Total	378	28	1,346

[a]Percentages calculated only for scientists who answered each pair of questions.
SOURCE: NRC/AAAS Survey; data tabulations by staff.

These data also raise the question of what other areas of research, beyond the seven categories specified by the NSABB, the life scientists who took part in the survey felt were dual use research. Like the Fink committee and the NSABB, the committee recognized that the range of life sciences research that could raise dual use concern is potentially greater than that relating directly to microbial threats, which are the focus of both lists. (Another report, *Globalization, Biosecurity, and the Future of the Life Sciences* (NRC 2006a), addresses this issue directly.) Both the Fink committee and the NSABB also focused on laboratory research. Analysis of the potential threat of a bioterror attack on the U.S. milk supply using botulinum toxin (Wein and Liu 2005), would not fit into any of the seven categories as defined by the Fink committee or the NSABB. Likewise, certain agricultural or pharmaceutical research may be dual use from the standpoint of formulation or dissemination, yet not involve modifying any of the biological properties of an agent in ways that are described by the NSABB or Fink report seven categories.

Hence, other categories of research having dual use potential clearly

exist, and therefore the possibility that some survey participants considered the theoretical potential of their research as dual use cannot be excluded. More exploration would be needed to understand the basis for the answers to these questions and more clarification will likely be needed on the part of the NSABB and others to define the meaning and scope of dual use research of concern.

The next step was to disaggregate the scientists who worked with dual use research, the seven categories of experiments, and select agents to try to gain insight into whether there were scientists working in different employment sectors who were most involved in this research. Table 3-4 lists the percentage of respondents for each employment sector who classify their current research as having dual use potential, one of the seven categories of experiments, or a select agent. In all three cases, government scientists were more likely to answer affirmatively. Academics answered similarly to industry scientists except in the case of dual use research, where life scientists in academia were much less likely to report that they were doing dual use research compared with scientists in all other sectors.

Summary of Key Results

- Sixteen percent of life scientists who responded to the survey considered their current research to have dual use potential. Five percent of these life scientists reported that they were currently working on research involving the seven categories of experiments outlined by the NSABB. Twenty-five percent of the respondents had worked with select agents at some point in their careers (Table 3-1).
- Of the three types of research outlined in the preceding item, the relationship among the types is complex. Two percent of scientists who responded to the survey had conducted all three types of research: dual use, one of seven categories of experiments, and work with select agents. Sixty-four percent of the research of these life scientists involved none of these (Table 3-2).
- Among the respondents who indicated that their research involved one of the seven categories of experiments, 30 percent of them did not consider their research to be dual use (Table 3-3).
- Government scientists who responded to the survey were most likely to report that they have done research involving dual use, the seven categories of experiments, or select agents. Academic scientists who responded to the survey were less likely to say that they were doing research they considered to have dual use potential than any other type of life scientist who responded to the survey (Table 3-4).

SCIENTISTS' VIEWS ON BIOTERRORISM

Background

The survey sought to ascertain the views of members of the life sciences community who participated in the survey about bioterrorism because the committee hypothesized that those attitudes, in particular perceptions of the risk of bioterrorism, might affect attitudes toward dual use research and what aspects of research are most likely to contribute to bioterrorism. Attitudes about bioterrorism can be assessed in many ways. For example, there are several possible reference points, including the perceived risk relative to other types of terrorist attacks or other domestic or international risks (e.g., natural disasters, pandemics, climate change, economic downturn, and political violence). Moreover, one can focus on the likelihood of an attack versus the severity of such an attack. Finally, different groups of people, for instance, the general public, the scientific community, or those with specific expertise in either terrorism or biological weapons, might have different views.

Other surveys exemplify the different approaches that have been taken to assess views about bioterrorism. For example, a survey conducted in June 2002, shortly after September 11 and the subsequent anthrax letters, found that the public thought bioterrorism was a serious threat to the quality of life in the United States: 74 percent said "Very Serious," 17 percent said "Somewhat Serious," 4 percent said "Not Too Serious," 1 percent said "Not Serious at All," and 3 percent responded "Don't Know/Refused" (Kaiser Family Foundation 2002). This indicated that in the aftermath of the terrorist attacks, the public was quite concerned about the potential for another bioterrorism attack.

A subsequent survey conducted by the Council for Excellence in Government in February 2004 found that a bioterrorism attack was the most frequently cited by the public as the "type of possible terrorist attack that most worries me": 48 percent said a bioterrorism attack, 37 percent said a chemical weapons attack, 23 percent said a nuclear attack, 21 percent said a suicide bomber, 13 percent said a plane hijacking, 9 percent said a cyberterrorism attack, and 4 percent said they were not sure (Council for Excellence in Government 2004:11).[13]

A third survey about the nature of the biological weapons threat, focusing on individuals involved in U.S. policy making and conducted during October–November, 2006 by the Center for Strategic and International Studies (CSIS) reported in part:

[13]Note that respondents could select multiple choices.

- "Biological weapons are a major threat that is viewed as somewhat increasing, greater than chemical weapons threat and, by a slim majority, a threat greater than or equal to the threat of nuclear weapons.
- The top two reasons for an increasing bioweapons threat are the increasing availability of dual use know-how, technology, and equipment and the revolution in the life sciences creating technologies and know-how that makes biological weapons acquisition easier.
- A major biological attack is somewhat unlikely within five years and somewhat likely or, according to over a quarter of those surveyed, very likely within a decade" (CSIS 2006).

Finally, *Foreign Policy Magazine*, in cooperation with the Center for American Progress, conducted three surveys (one in 2006 and two in 2007) to create its Terrorism Index; approximately the same group of 100 foreign policy experts was surveyed each time.[14] In 2006, after asking about the likelihood of a terrorist attack on the scale of September 11 or the attacks in London and Madrid within the next 5 to 10 years, the *Foreign Policy Magazine* survey asked, "Regardless of what you think about the timing of an attack, what two methods are most likely to be used in America by global terrorists?" Only 9 percent listed a biological weapons attack and 6 percent a nuclear weapon attack while 67 percent of respondents listed a suicide bombing attack and 66 percent listed an attack on major infrastructure. The February 2007 survey did not include questions related to biological terrorism and the September 2007 survey asked a different question from the 2006 survey: "In your view, what is the *single greatest* threat to U.S. national security?" Nuclear weapons and nuclear materials were considered the greatest threat (26 percent of respondents), followed by "Islamicism/Al Qaeda/Jihadists" at 20 percent. Biological terrorism was considered the single greatest threat by 2 percent of respondents.

These previous surveys did not directly focus on ascertaining scientists' opinions about bioterrorism or their views about the risks posed by advances in research. The current survey allows us to begin to examine whether life scientists, who are technical experts and who may be involved in the management of risk, share the same perceptions of risk as others who drive the formation of public policy. The information may also be useful input for instituting effective educational outreach to scientists.

Results and Discussion

The first set of questions in the survey focused on views about the likelihood of a bioterror attack. Three questions were posed:

[14]Available at http://www.foreignpolicy.com/story/cms.php?story_id=221#2.

• What is the percent chance that an act of bioterrorism will occur somewhere in the world in the next 5 years?

• What is the percent chance that an act of bioterrorism will occur in the United States in the next 5 years?

• What is the percent chance that knowledge, tools, or techniques from dual use life sciences research will facilitate an act of bioterrorism in the next 5 years?

In all three cases, life scientists who responded to the survey reported percentages ranging from 0 to 100 percent, thus indicating a divergence of views. Figures 3-1 to 3-3 present frequencies of responses to these questions.

A total of 1,587 respondents answered the three questions. On average, respondents thought that there was a 51 percent chance that there would be an act of bioterrorism somewhere in the world in the next 5 years (Figure 3-1); a 35 percent chance that there would be an act of bioterrorism in the United States in the next 5 years (Figure 3-2); and a 28 percent chance that dual use research would facilitate an act of bioterrorism (Figure 3-3). However, the three figures are not normally distributed. For the first two, the most frequent response is "50 percent" and in the first

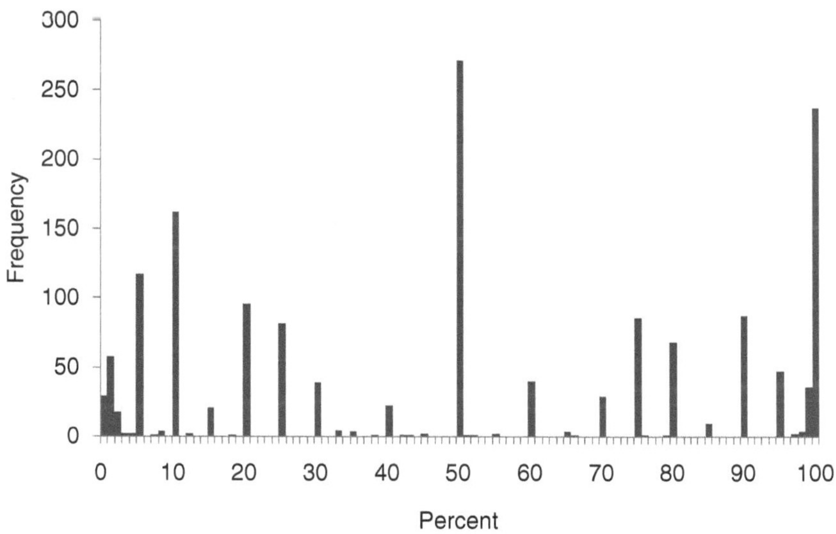

FIGURE 3-1 Frequency distribution for percentage chance of bioterror attack somewhere in the world.
SOURCE: NRC/AAAS Survey; data tabulations by staff.

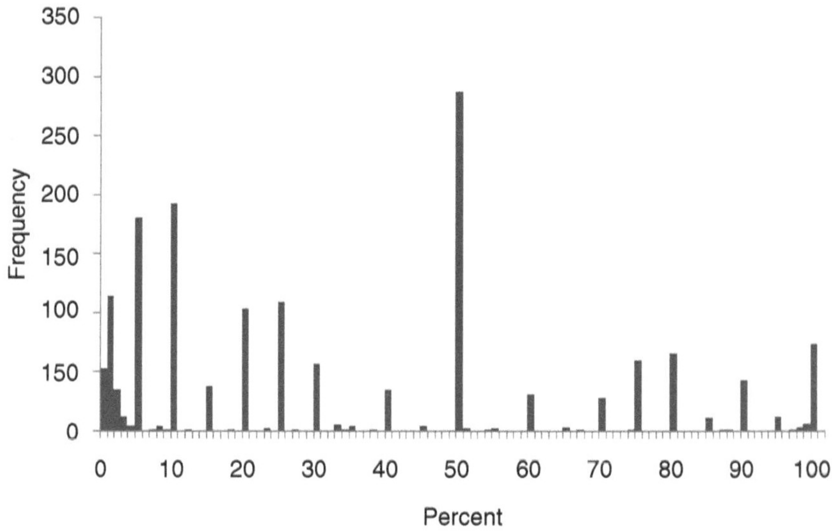

FIGURE 3-2 Frequency distribution for percentage chance of bioterror attack in the United States.
SOURCE: NRC/AAAS Survey; data tabulations by staff.

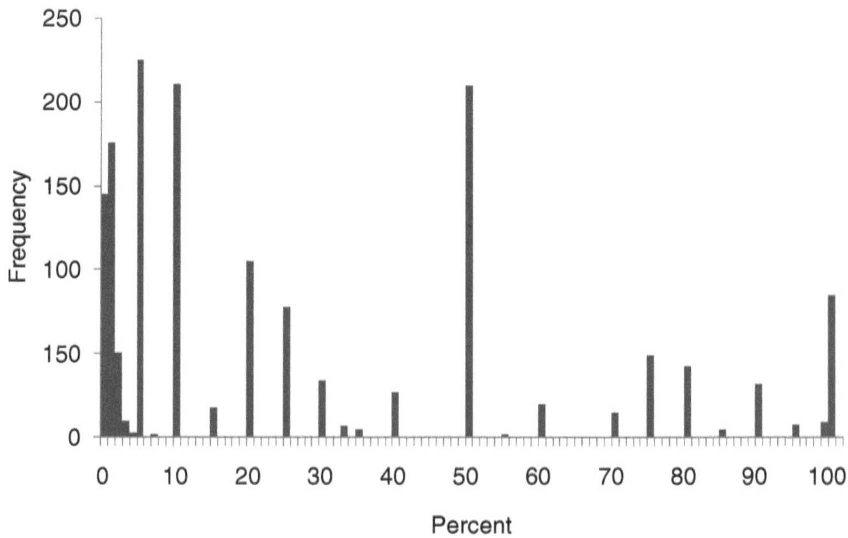

FIGURE 3-3 Frequency distribution for percentage chance of dual use life sciences research facilitating a bioterror attack.
SOURCE: NRC/AAAS Survey; data tabulations by staff.

figure, the second most frequent response was that the percentage chance of a bioterror attack somewhere in the world was 100 percent. For Figure 3-3 the three most frequent answers, in order, were: 5 percent, 10 percent (also the median), and 50 percent.

Further analysis revealed that there was a strong positive correlation between views of the respondents on the likelihood of a bioterror attack in the world and in the United States ($r = 0.81$). There was also a positive relationship between their views about the role of dual use research in facilitating a bioterror attack and their views on the likelihood of a bioterror attack in the world ($r = 0.47$) and in the United States ($r = 0.49$). Part of the explanation for the high correlation between the likelihood of a bioterror attack in the world and in the United States was the assumption that individual respondents would view the likelihood of a bioterror attack somewhere in the world as larger than or equal to the likelihood of a bioterror attack in the United States, because the United States is part of the world. If a respondent thought the former was low, he or she ought to also think the latter is low (although if the former were seen as high, it is less clear how a respondent would view the latter).[15]

As noted above, there is a positive relationship between a respondent's views regarding the likelihood of a bioterror attack occurring and his or her views of the likelihood of dual use research facilitating a bioterror attack. Additionally, a respondent's research might affect his or her view of whether dual use research could facilitate a bioterror attack. Hypothetically, respondents who work with select agents or dual use research might believe that there is a lower chance of dual use research facilitating a bioterror attack because these scientists do not see this as a serious risk relative to other potential types of bioterrorism involving less sophisticated agents. Conversely, respondents may see a higher risk because they understand how easy it would be to conduct an attack. Appendix Table D-2 compares the average perceived percentage chance that dual use research will facilitate a bioterror attack for respondents who do and do not consider their work to be (1) dual use or (2) within the seven categories of experiments, or (3) work with select agents. As noted previously, for all respondents, the perceived chance was an average of 28 percent. Each of the groups of respondents was essentially the same; that is, the perceived percentage chance for any group was within 3 percent (plus or minus) of the general average. The type of research

[15]An examination of the data showed that in 1,029 cases, respondents wrote in a larger percentage for an attack occurring somewhere in the world than for an attack occurring in the United States. The size of the percentages was reversed in only 29 cases. In the remainder of cases, the percentage chance of a bioterror attack was viewed to be the same in the world and in the United States.

that the respondents were engaged in did not appear to influence their views regarding the likelihood of dual use research facilitating a bioterror attack.

The survey then asked a question about the reasons why there have been so few acts of bioterrorism. The question in the survey was: "To date, there have been few acts of bioterrorism. Which of the following help explain why?" Respondents were given five possible explanations and asked with which they agreed:

- Terrorists lack the knowledge to work with or create dangerous biological agents.
- Terrorists lack the equipment to work with or create dangerous biological agents.
- Terrorists lack access to dangerous biological agents.
- Terrorists are deterred by the threat of being caught and punished.
- Terrorists prefer to use other means.

Figure 3-4 sorts the answers to the five possible explanations, with the highest percentages of "Yes" answers at the top of the chart.

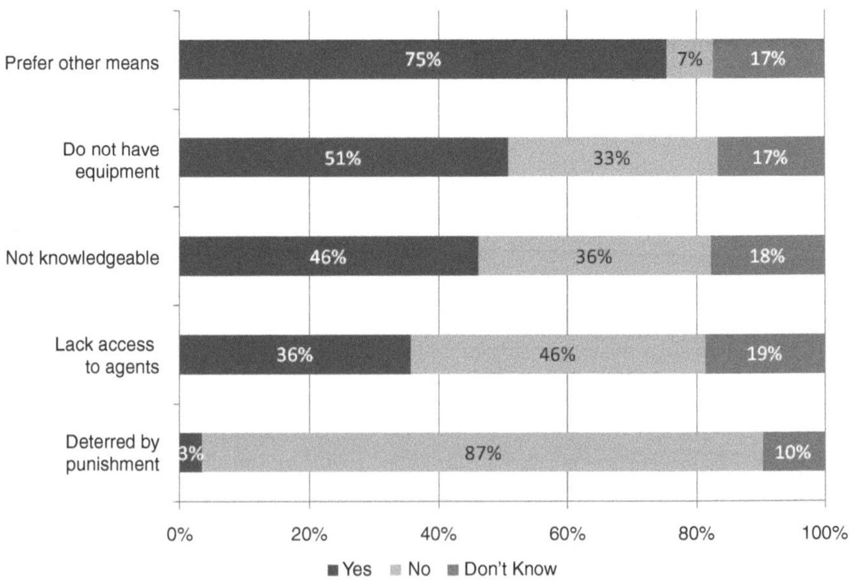

FIGURE 3-4 Reasons why there have been few acts of bioterrorism.
NOTE: Based on 1,588 respondents.
SOURCE: NRC/AAAS Survey; data tabulations by staff.

As Figure 3-4 indicates, 75 percent of the respondents believe that terrorists prefer other means of attack to the use of biological materials. An even greater majority (87 percent) believe that terrorists are not deterred by the threat of punishment. Just over half of the scientists (51 percent) felt that lack of equipment was a factor limiting bioterrorist attacks. Almost half of the life scientists (46 percent) felt that terrorists lacked the knowledge necessary to create a biological agent and carry out a biological attack. Only 36 percent thought that lack of access to biological agents that could be used in a bioterrorist attack was a major reason limiting such attacks, whereas 46 percent felt that access to biological agents was not a critical factor and 19 percent were uncertain. The fact that there is not a strong consensus that limiting access to pathogens and toxins will deter bioterrorism is an interesting result given that the focus of much counter-bioterrorism policy is on increasing security over dangerous pathogens (see, for example, the recommendations of the Commission on the Prevention of WMD Proliferation and Terrorism [2008]).[16]

The findings above may reflect the current situation, in which attacks with conventional weapons predominate, as well as the respondents' judgments about terrorists' knowledge and access to biological agents or equipment. As discussed in Chapter 1, it may also reflect a continuing debate over whether the most likely bioterrorist attack would be one that uses little advanced technology, as has been the case historically, and whether that could change in the future through the misuse of beneficial advancing biotechnologies (Frerichs et al. 2004; NRC 2004a, 2006a; Leitenberg 2005).[17] Box 3-1 contains some examples of comments on bioterror-

---

[16]The Commission on the Prevention of WMD Proliferation and Terrorism (2008) reported that "unless the world community acts decisively and with great urgency, it is more likely than not that a weapon of mass destruction will be used in a terrorist attack somewhere in the world by the end of 2013. The Commission further believes that terrorists are more likely to be able to obtain and use a biological weapon than a nuclear weapon." Among its recommendations the Commission said that "The United States should undertake a series of mutually reinforcing domestic measures to prevent bioterrorism" . . . including acting to "promote a culture of security awareness in the life sciences community." With regard to the dual use dilemma the Commission said that "the only way to rule out the harmful use of advances in biotechnology would be to stifle their beneficial applications as well—and that is not a realistic option. Instead, the dual use dilemma associated with the revolution in biology must be managed on an ongoing basis. As long as rapid innovations in biological science and the malevolent intentions of terrorists and proliferators continue on trajectories that are likely to intersect sooner of later, the risk that biological weapons pose to humanity must not be minimized or ignored."

[17]To date, the known cases of bioterrorism have used relatively unsophisticated approaches, as in 1984 when followers of the Indian guru Bhagwan Shree Rajneesh contaminated salad bars at 10 restaurants in The Dalles, Oregon, with *Salmonella* bacteria and sickened about 750 people (Carus 2000). Publicly available information also suggests that attempts at more sophisticated bioterrorism, such as the research conducted by scientists working with

BOX 3-1
Illustrative Respondent Comments on Why There Have Been Few Acts of Bioterrorism

"The information on creating a biohazard already exists in the biologic literature, the internet, college textbooks. The lack of use of these biohazards in terrorism is probably due to the lack of scientific education or the satisfaction of using explosives rather than biohazards."

"While it is important to have policies regarding oversight of dual use research, the reality is that 'terrorists' that are ideologically driven can come from any walk of life including trained PhDs."

"All good tools that help mankind can be misused to invoke terror. Dual use is a moral choice. Science should not be hampered because there are evil people in the world. Evil people will always find tools for terror."

ism provided by respondents at the end of the survey in which they were invited to offer open-ended comments on any biosecurity topic. Several hundred comments were offered, and a few examples are provided in this chapter as illustrations; they should not be considered in any way representative of the overall views of respondents.

Finally, the survey asked about the sources of information for individuals who might want to cause harm with a biological agent. The overarching question was "Do the following means of communication provide sufficient information for an individual with college-level life science training to deliberately create a harmful biological agent?" The respondents were asked to choose from the Internet, scientific journal articles, personal communication, and presentations.

Figure 3-5 summarizes the responses. Half of the life scientists who responded to the survey felt that the Internet could provide sufficient information to be used to deliberately create a harmful biological agent. Forty percent of these life scientists thought the information contained in journal articles would provide sufficient information. A similar proportion of life scientists who responded to the survey thought personal communication could provide enough information. Only 18 percent of the respondents thought that material presented at conferences would convey

the Japanese terrorist organization Aum Shinrikyo, have not been successful (Leitenberg 1999). The U.S. government has concluded that an American microbiologist working at the U.S. Army Research Institute for Infectious Diseases was responsible for the 2001 anthrax attacks, although the level of sophistication involved in the attacks was not high (FBI 2008).

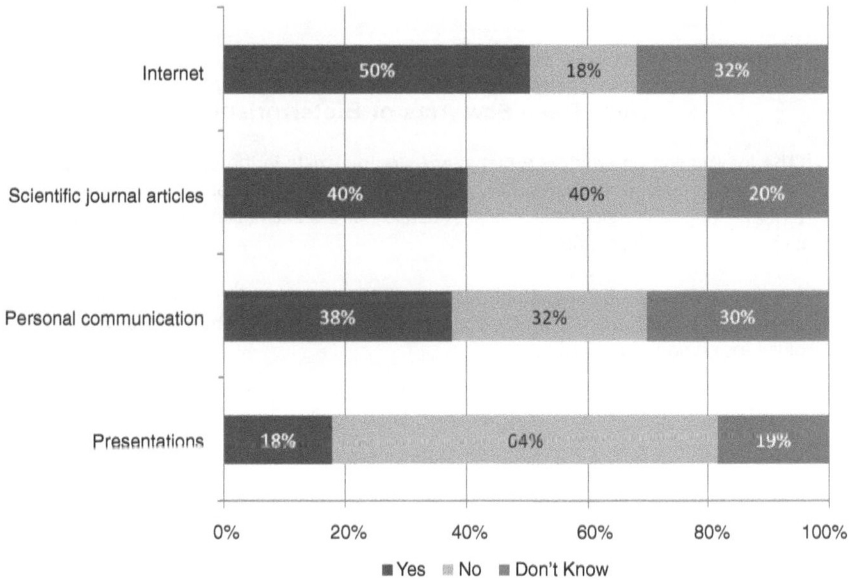

FIGURE 3-5 Respondents' views on whether sources of information could provide sufficient information for an individual with college-level life science training to deliberately create a harmful biological agent.
NOTE: Based on 1,588 respondents.
SOURCE: NRC/AAAS Survey; data tabulations by staff.

BOX 3-2
Illustrative Respondent Comments About Personal Communication and Scientific Journals Providing Information

"Certain research endeavors, including constructing pathogens from genetic fragments, or 'resurrecting' extinct pathogens and providing public access to sequence data from such experiments are of dubious scientific importance and journals, as well as funding agencies, should severely restrict such projects unless especially convincing justification can be provided."

"Increasing restrictions on publication of dual use research would be the most effective way of decreasing the amount of dual use research that is performed. That would have the effect of decreasing the possible threats from such research (as well as the benefits)."

"If scientific research is restricted significantly, then we lose in kind the ability of the scientific community to respond to biological emergencies and imperatives. If scientific collegiality shifts toward scientific nondisclosure, everyone loses."

enough information to allow an individual with college-level life science training to deliberately create a harmful biological agent compared to 64 percent who disagreed and 19 percent who were uncertain.

There were differing opinions among those responding to the survey regarding concerns about personal communication and scientific journals providing information that could be misused, as illustrated by the comments in Box 3-2.

Summary of Key Results

- On average, life scientists who responded to the survey felt that there was a 51 percent chance that an act of bioterrorism would occur somewhere in the world in the next 5 years and a 35 percent chance that an act of bioterrorism would occur in the United States in the next 5 years (Figures 3-1, 3-2).
- On average, life scientists who responded to the survey felt that there was a 28 percent chance that dual use research would facilitate an act of bioterrorism (Figure 3-3).
- Analysis showed that views of life scientists who responded to the survey about the chances that dual use research would facilitate a bioterror attack were not affected by the type of research they conducted (Appendix Table D-2).
- Life scientists who responded to the survey indicated that they think the reason that there have been few bioterror attacks to date is because terrorists prefer other means (e.g., conventional explosives) and not because they are deterred by punishment (Figure 3-4).
- The scientists who responded to the survey were more likely to agree that a reason for the low frequency of bioterror attacks was that terrorists lack the knowledge or equipment (46 and 51 percent, respectively); only 36 percent thought lack of access to the agents themselves played a contributing role, indicating that there was not a strong consensus that limiting access to pathogens and toxins would be a strong deterrent to bioterrorism (Figure 3-4).
- When asked to choose among sources of information that could potentially enable the creation of a dangerous biological agent, half of the scientists who responded to the survey thought that the Internet would be the most useful source for someone with college-level science training (Figure 3-5).

CHANGES IN BEHAVIOR IN RESPONSE TO DUAL USE CONCERNS

Background

One topic of interest to the committee was whether life scientists are already making changes in the conduct of their research and communication as a result of potential risks of bioterrorism. There are reports of scientists choosing to work in different fields or destroying collections of microorganisms in response to the strictures of the select agent rules (Schemo 2002; Fischer 2006). Fischer (2006) states that anecdotal reports and interviews suggest that an uncounted number of individual scientists in both academic and government laboratories destroyed pathogen collections prior to the initial select agent registration process. She found that: "All 28 scientists who responded to the Stimson survey on select agent regulations indicated that they had eliminated at least one research project involving a select agent in response to the new biosecurity regulations, and that they personally knew colleagues who had done the same" (Fischer 2006:30). She also reports that all respondents had eliminated or changed one or more of their international collaborations in response to the select agent regulations. With specific reference to dual use research of concern, at least one project proposal submitted to the National Institutes of Health (NIH) was reportedly never pursued because the study section raised dual use concerns; the investigators withdrew the proposal (Davidson et al. 2007). On the other hand, the only statistic available regarding changes in research communication in response to dual use research of concern is the widely cited figure that, of the 16,000-plus manuscripts reviewed by the 11 journals of the American Society of Microbiology (ASM) since it began screening manuscripts for dual use research, only 4 manuscripts had actually been modified in any way (these were published with minor modifications) and only 2 others were not published because the authors were unwilling to provide full methodological details (Kaplan 2008).[18]

In an attempt to gain some empirical evidence about potential changes in behavior, the survey asked: "Have you made any changes in how you conduct or manage research because of concerns that knowledge, tools, or techniques from your research might be deliberately misused to facilitate

[18]Publishing a paper without full methods would mean it could not be repeated—the ASM considers that not providing full methodological details would undermine science; the position that science is too important to be undermined in this manner was supported by participants in a January 2003 workshop organized by the National Academy of Sciences and the CSIS (see Chapter 1) and by the Journal Editors and Authors Group (2003a,b,c, Fox 2003).

bioterrorism?" Nine possible changes were offered on the survey, covering the scientific process from project origination to journal submission:

- I decided against conducting a specific research project/experiment.
- I decided to shift my research away from an area altogether.
- I decided against seeking funding for a proposed research project.
- I decided against collaborating with particular scientists, postdocs, students, etc.
- I limited my conversations about my research.
- I decided against submitting a manuscript to a journal.
- I modified a manuscript.
- I decided against presenting research at a conference.
- I modified a conference presentation.

As discussed further below, most members of the committee were surprised by the number of respondents who reported changing how they conduct and communicate their research, and how they think about collaborations because of dual use concerns.

Results and Discussion

As a first step in the analysis, the nine possible changes were aggregated to determine whether *any* change in research behavior had been made in response to dual use concerns. Fifteen percent (260 out of 1,744) of the respondents who answered this question said they had made one or more changes in their research activities in response to dual use concerns. As noted above, this was a finding that most members of the committee did not expect.[19]

A number of factors may explain why some scientists have made a

[19]An informal poll was taken among attendees at the 2nd International Forum on Biosecurity in Budapest, Hungary, in March 2008 to gain anecdotal evidence as to how some members of the biosecurity community viewed the likelihood that scientists in the United States were changing research behaviors in response to dual use concerns. The attendees, most of whom were life scientists, were asked to indicate via a show of hands, "How many members of the U.S. life sciences community have taken actions to protect dual use research of concern?" These actions included not doing research, altering collaborations, not presenting data at a meeting, and modifying manuscripts. The results for the 59 attendees who took part were:

Very Few (<10 had made changes) — 43 (73%)
Some (approximately 100) — 15 (25%)
Many (>250) — 1 (2%)
Very Many (>1,000) — 0 (0%).

TABLE 3-5 Percentage of Respondents Who Made a Change to Their Research on the Basis of Type of Research and Employment Type

	Made a Change (%)		
	No	Yes	Observations[a]
Dual use?			
No	89	11	1,153
Yes	62	38	213
Seven categories of experiments?			
No	86	14	1,284
Yes	66	34	82
Works with select agents?			
No	89	11	1,119
Yes	76	24	453
Employment type			
Industry	81	19	204
Academia	88	12	1,002
Government	79	21	121
Other	91	9	65

[a]Applies to respondents who answered pairs of questions (e.g., answered whether the respondent made a change and whether the respondent considered his or her research dual use.

SOURCE: NRC/AAAS Survey; data tabulations by staff.

change, while others have not. Factors may include a scientist's beliefs about the responsible conduct of research or the nature of biosecurity threats on the one hand and the nature of his or her research on the other. Variables that might be associated with changing behavior and for which data were collected in the survey included whether a scientist's research had dual use potential, worked with select agents,[20] or could be catego-

A large majority (73 percent) of the participants in this informal poll predicted that fewer than 10 scientists would have modified their behavior due to dual use concerns. Approximately 25 percent of those polled believed that approximately 100 scientists would have modified their behavior. In comparison to this study, the group in Budapest clearly underestimated the number of U.S. life scientists who reported that they have changed their research behavior in response to dual use concerns.

[20]The select agent regulations prohibit foreign nationals from countries designated as supporting terrorism from being given access to select agents within the United States. Also, NIH has restricted international collaboration on select agent research unless the foreign

rized as one of the seven categories of experiments defined above, and employment sector.

To test whether there was a relationship between a respondent's type of research or employment and whether he or she had made a change, three crosstabs were used, as shown in Table 3-5. As the table shows, respondents who worked with dual use research, the seven categories of experiments, or select agents were at least twice as likely to make a change in their research. Those who said they considered their research to be dual use were 3.5 times more likely to say they had changed their research; those who considered their research to involve the seven categories of experiments identified by the NSABB were 2.4 times more likely to have changed their research; and those who had worked with select agents were only 2.2 times more likely. While work with select agents increased the likelihood of changing research, unlike the findings of Fischer (2006) discussed earlier, only 24 percent of the scientists who responded to the survey reported having worked with select agents also reported changing their research. Industry and government respondents were more likely to make a change in their research than were academic and other respondents.

A more detailed look at the behavioral changes that were made indicates that many of the actions taken by the life scientists who participated in the survey involved changes in the early stages of research and initial communication of research results prior to submission of a manuscript to a scientific journal (Table 3-6). The table shows that the types of behavioral changes reported by respondents.

As Table 3-6 shows, the *most likely* change in conduct or management of research was to limit conversation about the research (9 percent), followed by deciding against collaborating with particular scientists, postdoctoral fellows, or students (4 percent) and deciding against conducting a specific research project or experiment (4 percent). Relatively few reported modifying a presentation at a conference (2 percent). The *least likely* change was to decide against submitting a manuscript for publication (1 percent). Only 2 percent of the respondents indicated that they had modified a manuscript. But even this number is substantially higher than the number commonly cited based upon the experience of the ASM journals.

The committee found it disturbing that a few respondents expressed negative views toward foreign scientists, as exemplified in Box 3-3.

A correlation matrix (Table 3-7) was created to see if scientists who made one type of change, also made others. Making one change was

laboratories meet stringent biosafety and biosecurity standards (see http://www.niaid.nih.gov/ncn/grants/selectterm.htm).

TABLE 3-6 Types of Changes Scientists Made Because of Concerns That Knowledge, Tools, or Techniques from Their Research Might Be Deliberately Misused to Facilitate Bioterrorism

Change Made	Frequency	Percentage
Change in performance of research		
Decided against conducting a specific research project/experiment	69	4
Decided against seeking funding for a proposed research project	58	3
Decided to shift my research away from an area altogether	49	3
Change in collaboration		
Decided against collaborating with particular scientists, postdocs, students, etc.	72	4
Change in research communication		
Limited conversations about research	156	9
Decided against presenting research at a conference	31	2
Modified a conference presentation	40	2
Decided against submitting a manuscript to a journal	24	1
Modified a manuscript	38	2
One or more changes	260	15
Did not make change	1,484	85
Respondents	1,744	—
Did not answer	210	—
Total	1,954	—

NOTES: Based on 1,744 respondents. An individual respondent could select more than one category. Some respondents made only one or a few changes.
SOURCE: NRC/AAAS Survey; data tabulations by staff.

positively associated with making others, such as deciding not to present research and not submitting a manuscript for publication, or modifying both a presentation and a manuscript. Additionally, respondents who made changes often made multiple ones, as shown in Table 3-8.

The results of the survey indicate that more of the respondents to the survey than expected have taken action on security concerns regarding release of dual use information well before publication by limiting their conversations with other scientists and in scientific conference presentations. Clearly these life scientists are sufficiently concerned that their research can be misused that they are acting to limit the risk that dual use research of concern could contribute to bioterrorism. The reported limits on communication and collaborations raise the question of whether dual use research security concerns may be having an impact on the traditional openness that characterizes the conduct of the life sciences.

TABLE 3-7 Correlations Between Changes Scientists Made

Variable	Against Conducting a Project	Shift Research Away From an Area	Not to Seek Funding	Not to Collaborate	Limited Conversations	Not to Submit Manuscript	Modified a Manuscript	Not to Present Research	Modified a Presentation
Against conducting a project	1.00								
Shift research away from an area	0.37	1.00							
Not to seek funding	0.41	0.45	1.00						
Not to collaborate	0.30	0.28	0.36	1.00					
Limited conversations	0.26	0.26	0.24	0.37	1.00				
Not to submit manuscript	0.25	0.34	0.28	0.37	0.33	1.00			
Modified a manuscript	0.21	0.24	0.26	0.28	0.28	0.35	1.00		
Not to present research	0.24	0.37	0.29	0.32	0.43	0.65	0.40	1.00	
Modified a presentation	0.17	0.23	0.21	0.33	0.41	0.31	0.58	0.50	1.00

NOTES: This uses phi and is based on 1,744 respondents.
SOURCE: NRC/AAAS Survey; data tabulations by staff.

BOX 3-3
Illustrative Respondent Comments About Foreign Scientists

"Federal funds should not be used to support dual use research outside of the United States, unless it is through existing programs specifically designed to "re-train" former biological weapons scientists, and whose programs are well monitored for both fiscal responsibility and scientific progress."

"It is not in US interest to train nationals of some nations in dual use technology."

"The largest threat is allowing foreign students and researchers access to all of our academic research labs and knowledge. Much of this knowledge gets exported to scientists' home countries through personal communications that are not detected by monitoring scientific publications and meetings. Many countries develop relationships with their scientists that work and live in the US specifically to get access to this information. A mechanism of the US government to work with US scientists to detect and monitor these kinds of activities, as well as greater restrictions to entrance to US institutions by students from certain countries and/or their participation in certain programs of study would be of most benefit in protecting the US from bioterrorism and other losses of security and economic interest."

"Federal funds should not be used to support dual use research outside of the United States, unless it is through existing programs specifically designed to "re-train" former biological weapons scientists, and whose programs are well monitored for both fiscal responsibility and scientific progress."

TABLE 3-8 Number of Changes Individual Scientists Made

Number of Changes	Frequency	Percentage
0	1,484	85
1	146	8
2	53	3
3	20	1
4	19	1
5	5	<1
6	7	<1
7	3	<1
8	2	<1
9	5	<1
	Number	Percentage
Respondents	1,744	100
Did not answer	210	—
Total	1,954	—

SOURCE: NRC/AAAS Survey; data tabulations by staff.

Summary of Key Results

- Fifteen percent of the respondents made one or more changes in their research activities in response to dual use concerns (Table 3-6).
- Life scientists responding to the survey who work with dual use research or select agents were more likely to make a change (Table 3-5).
- Respondents indicated that they have made changes throughout the scientific process. The most frequent change was to limit conversation with others. The least frequent was to not submit a manuscript for publication (Table 3-5).
- Respondents who made changes often made multiple changes. There was also a positive correlation between changes: Scientists making one type of change were more likely to make other changes as well (Table 3-7, 3-8).

RESPONSIBILITY FOR OVERSIGHT

Background

Debate intensified in late 2001 about who should have responsibility for oversight of dual use research and how the potential misuse of scientific research and knowledge in the life sciences should be mitigated. As discussed in greater detail in Chapter 1, some, such as the members of the Controlling Dangerous Pathogens Project at the University of Maryland, have argued that mandatory control at the international level is essential (Harris 2007; Steinbruner et al. 2007). Others have been strongly opposed to any government oversight, arguing that national security requires that scientific research move forward without any interference (Kaplan 2008). Another approach, supported by the Center for Biosecurity (Kwik et al. 2003) and the Fink committee (NRC 2004a), is self-governance and achieving national security objectives using a bottom-up approach that relies on the scientific community's own sense of responsibility, although the Fink committee supported an advisory role for the federal government in providing guidelines for appropriate behavior. The NSABB has put forward a proposal for a mixed approach that includes mandatory actions at the levels of research institutions and individual scientists (NSABB 2007). The comments from several respondents in Box 3-4 reflect the continuing debate over the locus, nature, and necessity of responsibility for oversight of research.

The survey examined the views of members of the scientific community regarding the allocation of responsibility, with specific questions about the responsibilities of individual scientists, journal editors, professional scientific societies, institutions, funding agencies, and the federal government. Each is examined in turn, with an initial brief description of

BOX 3-4
Illustrative Respondent Comments About Responsibility for Oversight of Research

"I strongly believe that dual use vigilance starts with the researcher and the editors of journals. Therefore extensive education of these groups about the issues and the best practices is urgent. I strongly believe that the role of government should be to primarily serve as an educational unit not a regulatory unit. Creating an advisory board that establishes and continually reviews best practices and educational goals is crucial. Oversight should happen at the individual institutional level similar to that of the IRB process with the advisory committee setting best practices for the individual institutional dual use committees."

"Regulations should be designed to encourage self-policing by institutions and principal investigators with some federal oversight. Comparable regulatory examples are radiation, cancerous agent, chemical mutagen, and animal handling."

"The Federal government should assure that basic standards used or met for dual use research that is based on consensus standards of the research areas. The government assures that the standards have teeth, the researcher communities evolve standards that are contextually appropriate and evolve as knowledge and conditions change."

the issue, followed by discussion and analysis of the results, with the key results listed again at the end.

Role of Individuals

The survey asked questions about whether life scientists, acting either individually or collectively as members of a scientific or professional society, can be responsible for biosecurity (self-governance). As discussed in Chapter 1, some measures have been proposed by which scientists could reduce the risk that their research might be misused. To encourage the development of a norm and sense of shared responsibility, some propose that scientists should take an oath similar to the Hippocratic oath that physicians take at graduation (Revill and Dando 2006). One proposal is that scientists should conduct an initial and continued review of research ideas to assess whether they have dual use potential. This is the approach that the NSABB has taken in its *Proposed Framework for the Oversight of Dual use Research* (NSABB 2007). Another approach would be to have scientists provide assurance to their employers that they are aware of whether their work has dual use potential. Yet another possibility is that

BOX 3-5
Illustrative Respondent Comments About the Role of Individuals

"Principal investigators need to be very careful about possible dual use of their research, and they can be assisted in this by grant reviewers. But formal review and regulation procedures almost certainly will hamper a great deal of innocent research without materially advancing our safety."

"Scientists need to understand the potential impact of their actions more than they need national regulations of those actions."

"There will always be a certain amount of risk associated with gaining knowledge about life. Scientists must allow ethics, rather than fear, to guide them in making responsible research decisions."

scientists should be responsible for training their students and colleagues about dual use issues. These approaches are not mutually exclusive, but each captures a slightly different idea about how individuals might be responsible for addressing dual use concerns. An illustration of some of these views is provided by the comments in Box 3-5.

Discussion

The survey included four questions to assess views about the responsibilities of individual scientists. The answers are displayed in Figure 3-6. Items with the highest percentages of "Strongly Agree" and "Agree" appear at the top of the chart. Almost 90 percent of the respondents felt that principal investigators (PIs) should be responsible for initial review of their research. A similarly high percentage supported PIs' taking responsibility for training their students about dual use concerns. Fewer than 40 percent supported the use of an oath. As discussed further in subsequent sections, support for voluntary responsible conduct is higher than for mandatory actions.

The committee examined the question of whether life scientists who responded to the survey answered these four questions similarly—that is, did they uniformly show support for or opposition to individual responsibility. Spearman's ρ was used between pairs of the variables as seen in Table 3-9. There is a positive correlation between each of the four individual responsibility variables, which suggests that respondents who sup-

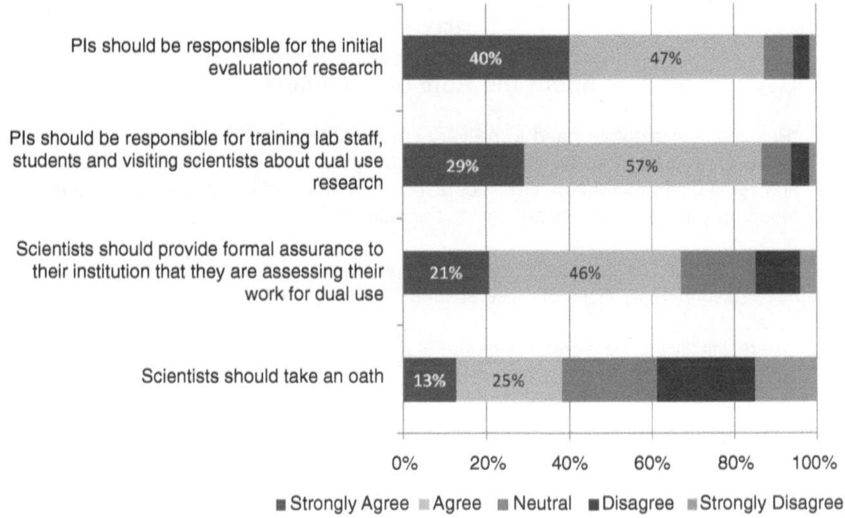

FIGURE 3-6 Respondents' views on individual responsibility.
NOTE: Based on 1,658 respondents, except for "PIs should be responsible for training lab staff," which is based on 1,637 respondents.
SOURCE: NRC/AAAS Survey; data tabulations by staff.

TABLE 3-9 Correlation Between Variables Relating to Respondents' Views on Individual Responsibility

	Variable			
Variable	PIs Should Conduct Initial Evaluation	Scientists Should Provide Formal Assurance	Scientists Should Take an Oath	PIs Should Train Staff, Students, Etc.
PIs should conduct initial evaluation	1			
Scientists should provide formal assurance	0.44 (1,658)	1		
Scientists should take an oath	0.11 (1,658)	0.4 (1,658)	1	
PIs should train staff, students, etc.	0.45 (1,637)	0.37 (1,637)	0.16 (1,637)	1

NOTE: Number of respondents in parentheses.
SOURCE: NRC/AAAS Survey; data tabulations by staff.

BOX 3-6
Respondent Comments Illustrating Range of Views on Individual Responsibility

"It is extremely difficult to define what constitutes "dual-use"; many benign materials can become toxic if used improperly. Principal investigators should be responsible for taking general precautions and training, but should not be held responsible for any and all potential use. 'Potential use' is also extremely difficult to define."

"Some inadvertent dual use research will always be a product of research, but some up-front thought about how to report the results would be helpful. I could envision as part of a short course on research ethics for graduate students that dual use research would be a necessary topic."

"If we don't regulate ourselves and something bad happens, the government is going to forcibly do it."

ported one form of personal responsibility also supported the others. Box 3-6 offers some examples of comments from the respondents, illustrating the range of views among them.

Summary of Key Results

- Almost 90 percent of the respondents agreed that PIs should be responsible for an initial review of their research and for training their students about dual use concerns (Figure 3-6).
- Just under 40 percent of life scientists who responded to the survey agreed that scientists should take a Hippocratic-style oath (Figure 3-6).
- Life scientists who responded to the survey tended to support one or more of the other approaches if they supported one of the approaches to individual responsibility noted above (Table 3-9).

Role of Journal Editors

As discussed in Chapter 1, the question of whether to publish the results of certain experiments that appear to pose potential dual use risks has been at the center of debates over whether open scientific communication could provide useful information to terrorists. Beginning in 2003, following a statement by a group of journal editors and scientists (Fox 2003; Journal Editors and Authors Group 2003a,b,c), a number of prominent journals undertook policies to provide for review of manu-

scripts that potentially raised dual use concerns. The committee wanted to know whether (1) respondents were aware of these policies and (2) if they thought such efforts were appropriate. The survey asked:

- Should scientific journals have policies regarding publication of dual use research?
- Do the journal(s) in your field require reviewers to evaluate whether manuscripts include knowledge, tools and techniques with dual use potential?
  —All of the journals have a policy
  —Some of the journals have a policy
  —None of the journals have a policy
  —Don't know
- Do the journal(s) in your field require authors to disclose any research with dual use potential to editors upon submission of the manuscript?
  —All of the journals have a policy
  —Some of the journals have a policy
  —None of the journals have a policy
  —Don't know
- Have you ever contacted an editor because you felt that a manuscript you were reviewing contained knowledge, tools, or techniques that could pose a threat to national security?[21]

Discussion

In response to the question of whether journals should have policies on publication of dual use research, a majority (57 percent) of the 1,755 respondents who answered the question thought that they should, but even here the percentage of "Don't Know" responses (19 percent) is larger than for many other questions in the survey.[22]

Further results are presented in Figure 3-7. Among the respondents, 16 percent thought all or some of the journals in their field had a policy for reviewers to evaluate dual use potential. Fewer life scientists who responded to the survey (12 percent) thought all or some journals had a

[21]Respondents could answer "no" in two ways—either because they had not reviewed manuscripts or although they had, they had not contacted an editor.

[22]As an interesting aside, the committee wondered whether there might be a positive association between respondents' views about how useful scientific journals were in providing information to those wishing to conduct a bioterror attack and respondents' support for policies on dual use research for scientific journals. However, no relationship was found ($r = -0.07$; $n = 1,266$).

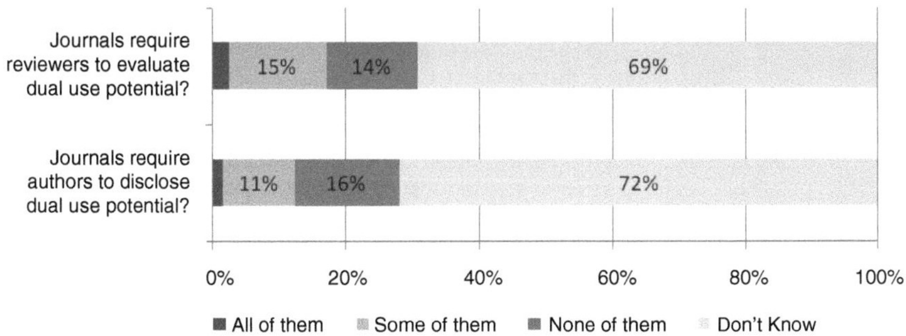

FIGURE 3-7 Respondents' views on whether journals require reviewers to evaluate, and authors to disclose, whether manuscripts include knowledge, tools, and techniques with dual use potential.
NOTE: Based on 1,755 respondents.
SOURCE: NRC/AAAS Survey; data tabulations by staff.

policy for instructing authors to disclose the dual use potential of their research when submitting a manuscript.

The very high percentage of respondents who answered "Don't Know" is striking. Even though a number of leading journals (ASM journals, Nature Publishing Group, *Science*, and *PNAS*) do have policies regarding review of dual use research, those policies may have not been effectively communicated to the scientific community. Another possibility is that most life scientists who belong to AAAS do not publish in the journals that have policies or have not had the occasion to learn about the policies because their work has not raised dual use concerns. For example, there is no stated biosecurity review policy for *Cell*, the highest impact journal specific to the life sciences. Or it may be that authors do not pay adequate attention to the instructions to authors. Box 3-7 contains some examples of the open-ended comments from respondents on these issues.

The survey also asked whether those who served as journal reviewers had contacted editors with a dual use concern. A total of 1,755 respondents answered the question. One percent of respondents said that they had contacted an editor because a manuscript they were reviewing contained knowledge, tools, or techniques that they deemed a potential threat to national security. Sixty-three percent said they had not, although they had reviewed manuscripts; the remaining scientists had not reviewed manuscripts.

BOX 3-7
Illustrative Respondent Comments on Journal Policies on Review of Dual Use Research

"Don't be paranoid, but don't be naive either. Many studies should be published that are potentially useful to a determined terrorist but not all; the degree of risk should be assessed. We should not curtail publication that would ultimately help in our defenses. Perhaps, a review board would need to be created for specific journals to advise authors of content that would be of significant harm; this has been done in other disciplines in the past."

"How can we possibly try to close the door on the free exchange of information at this point?"

"The sequence of the 1918 influenza virus should not have been made public. Although I prefer open publication of all scientific results, some information should be considered too sensitive for open dissemination. That sequence is a recipe for a future terror act."

Summary of Key Results

- A majority of life scientists (57 percent) thought journals should have policies on publication of dual use research.
- More than two-thirds of life scientists who responded to the survey, however, did not know whether journals in their field had policies to evaluate dual use potential (Figure 3-7).
- Sixteen percent of life scientists who responded to the survey thought all or some of the journals in their field had a policy for reviewers to evaluate dual use potential. Twelve percent thought journals had a policy for instructing authors to disclose research with dual use potential to editors when submitting a manuscript (Figure 3-7).
- One percent of respondents said that they had contacted an editor because a manuscript they were reviewing contained knowledge, tools, or techniques that they deemed a potential threat to national security.

Role of Professional Scientific Societies

The survey also asked several questions about views concerning the responsibilities of professional and scientific societies, which appear to be logical candidates for leading activities on addressing conduct of science and educating their members about their professional responsibilities. The Fink report, for example, recommended that "national and international

professional societies and related organizations and institutions create programs to educate scientists about the nature of the dual use dilemma in biotechnology and their responsibilities to mitigate its risks" (NRC 2004a:4). In recent years, as described in Chapter 1, much of the focus on professional societies has been on the potential contributions of codes of conduct for their members. The 2005 and 2008 intersessional meetings of the Biological Weapons Convention have discussed the contributions of codes of conduct for scientists in preventing the spread of biological weapons, with a number of professional societies invited to address plenary sessions during the meetings.[23] The survey thus asked scientists whether they believe that professional societies should have codes of conduct and whether the societies to which they belonged had a code.

Discussion

An overwhelming majority (82 percent) of the 1,743 respondents who answered the question felt that professional scientific societies should have codes for the responsible conduct of dual use life sciences research. Only 9 percent said "No"; the remainder responded "Don't Know." This was an especially interesting finding given the continuing controversy over whether codes are necessary or appropriate for scientists (Rappert 2004; Revill and Dando 2006, 2008; Macrina 2007; Kaplan 2008). (Additionally, there was a statistically significant positive relationship between support for professional societies having codes of conduct and support for an oath [Spearman's ρ = 0.18; n = 1,516]. Those who favored a code also tended to agree that scientists should take an oath.)

When asked whether they were members of any professional societies that already have codes of conduct that include statements about the responsible conduct of dual use research, however, most (66 percent) of the 1,743 respondents did not know. Sixteen percent said "Yes" and remainder said "No." In addition, of the 16 percent of the life scientists in the survey who said they were members of a society with a code of conduct that included dual use research, answers often were inaccurate. Table 3-10 provides a list of the most frequently cited professional societies that respondents believed had a code with a provision addressing dual use research.[24] The most frequently cited professional society, ASM, does have a code that addresses actions that might contribute to biological

[23]For documents related to the 2005 Meetings of Experts and States Parties, see http://www.opbw.org/, and for the 2008 meetings, see http://www.unog.ch/80256EE600585943/(httpPages)/F1CD974A1FDE4794C125731A0037D96D?OpenDocument.

[24]For codes of conduct of professional and scientific societies, see http://ethics.iit.edu/codes/codes_index.php, and the UN Web site cited in the preceding footnote.

TABLE 3-10 Professional Societies Most Frequently Cited as Having Codes of Conduct

Society	Frequency
American Society of Microbiology	101
American Association for the Advancement of Science	85
American Society for Biochemistry and Molecular Biology	12
American Chemical Society	11
Society for Neuroscience	10
American Association of Immunologists	7
American Phytopathological Society	5
Geological Society of America	5
New York Academy of Sciences	5

SOURCE: NRC/AAAS Survey; data tabulations by staff.

weapons or bioterrorism, although that code does not explicitly mention the issue of dual use research.[25] The next most frequently cited society, AAAS, does not have a code of conduct for its members.[26] In fact, no scientific society or professional scientific organization has a code specifically addressing dual use research, although some scientific societies, such as the ASM, American Chemical Society, and the American Phytopathological Society, have codes that cover biological or chemical weapons and preventing the misuse of science, which could include dual use research of concern even though they do not specifically use the term dual use.

Summary of Key Results

• Eighty-two percent of life scientists who responded to the survey felt that professional scientific societies should have codes for the responsible conduct of dual use life sciences research.

[25]"ASM members are obligated to discourage any use of microbiology contrary to the welfare of humankind, including the use of microbes as biological weapons. Bioterrorism violates the fundamental principles upon which the Society was founded and is abhorrent to the ASM and its members. ASM members will call to the attention of the public or the appropriate authorities misuses of microbiology or of information derived from microbiology." (http://www.asm.org/ASM/files/ccLibraryFiles/FILENAME/000000001596/ASMCodeofEthics05.pdf).

[26]In the late 1980s, AAAS formally decided against creating a broad code of conduct for their members. Instead, they pledged to help their affiliated societies develop codes of professional conduct specific for their members. For more information, see http://www.aaas.org/spp/sfrl/committees/csfr/. AAAS continues to issue statements on ethical conduct of scientists and to advocate for scientific integrity and responsibility. For examples of these statements, see *http://www.aaas.org/spp/sfrl/per/archives1.shtml*, http://www.aaas.org/spp/sfrl/per/newper, and http://fellowships.aaas.org/04_Become/04_Ethics_Policy.shtml.

• Many respondents to the survey (66 percent) did not know if they were a member of a society that had a code.

• Many scientists who indicated that the societies they belonged to had a code of conduct identified societies that did not have such a code.

Role for Institutions

Employers or home institutions are also potential loci for oversight. Although much of the responsibility under the NSABB's proposed framework for oversight of dual use research (NSABB 2007) would rest on individual scientists, NSABB also anticipates a role for institutions in areas such as reviewing proposals that investigators have identified as posing dual use concerns. Even for government employees, an individual agency might be perceived as a more acceptable choice than the generally defined "federal government." The survey asked two questions about whether institutions (as opposed to PIs in earlier questions) should provide education and training, and whether research institutions should review all grant proposals with dual use potential prior to submission.

Discussion

Survey participants were asked if they agreed with three statements pertaining to institutional involvement in education and oversight:

• University and college students should receive educational lectures and materials on dual use life sciences research including the potential that knowledge, tools, and techniques of such research could pose a threat to national security.

• Institutions should provide mandatory training for scientists regarding dual use life sciences research.

• All grant proposals for life sciences research with dual use potential should be reviewed by a researcher's institution prior to submission for funding.

The answers to these questions are shown in Figure 3-8, with the responses receiving the highest percentage of "Strongly Agree" and "Agree" listed at the top. The committee noticed in particular the high level of support (68 percent) for providing education about dual use issues to university and college students. The idea of mandatory training for scientists, which is part of the NSABB's proposed framework and a recommendation by the Commission on the Prevention of WMD Proliferation and Terrorism (2008), attracted a majority of support, and fewer than 10 percent strongly disagreed. The idea of a review for dual use potential was less

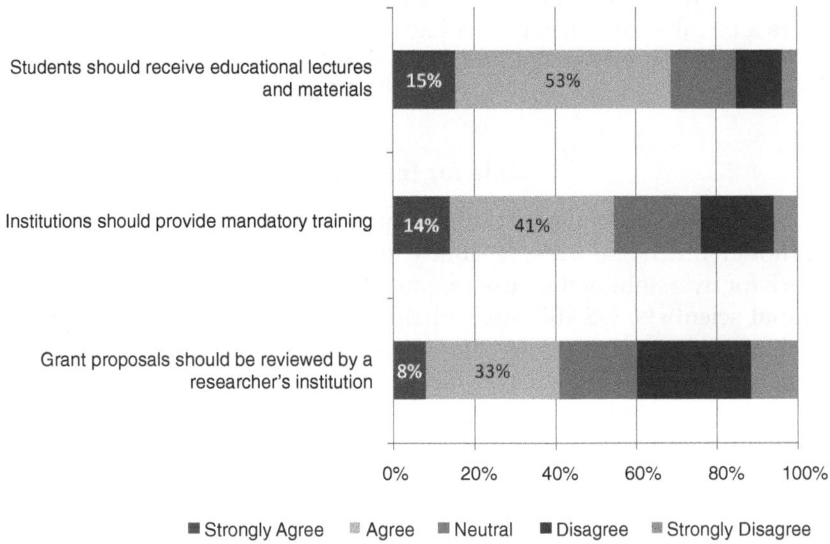

FIGURE 3-8 Support for education or training and review of grants by institutions.
NOTES: Based on 1,637 respondents, except for "Grant proposals . . ." which was based on 1,633 respondents.
SOURCE: NRC/AAAS Survey; data tabulations by staff.

popular, with agreement and disagreement split at 40 percent each and an estimated 20 percent neutral. This may reflect perceived problems with some of the current institutional review bodies (i.e., institutional biosafety committees, institutional review boards, and institutional animal care and use committees) or with uncertainty about how the review process would be implemented. Some illustrative open-ended comments can be found in Box 3-8.

Summary of Key Results

- There was a high level of support (68 percent) for providing education about dual use issues to university and college students (Figure 3-8).
- The idea of mandatory training for scientists also attracted a majority of support (Figure 3-8).

BOX 3-8
Illustrative Respondent Comments on Education and Training

"I think there are already useful restrictions on potentially harmful agents. The key from here on in is education and awareness."

"I am in favor of universities increasing their discussion of dual use research with faculty, staff and students and not increasing the amount of federal oversight."

"Dissemination and use of a handful of specific agents and reagents should be regulated. Experimental practices and dissemination of information re results should be managed by education regarding dual use dangers and institutional review (rather than government regulation)."

"Training is your primary deterrent. If people know and/or recognize potential hazards, usually they will report it if the atmosphere is not too detrimental or restrictive and heavy handed."

"Education, not regulation. Peer pressure versus governmental censorship and research restrictions. It is dual purpose: why block life-supporting research for potential misuse? All technology and definitive research is a double-edged sword: radiation to treat cancer and to kill people, molecular engineering to improve crop production and induce plagues. I am fed up with governmental paranoia. Yes, be concerned and watchful, but carry on normal life."

"Good training in ethics is essential for everyone in science. The other major thing that will prevent bioterrorism is to widely promulgate the contact information for responsible public authorities who can investigate potential threats or potentially dangerous individuals that scientists may identify."

Role for Funding Agencies

Funding agencies could play a role in oversight since they would see research proposals and progress reports. This could be considered part of the more general question about federal oversight addressed in the next section, since the majority of funding for life sciences research comes from the federal government. But we wanted to ask specifically about the role of funding agencies since at least the major funders in the United Kingdom have initiated a review for biosecurity concerns (BBSRC/MRC/Wellcome Trust 2005).

The survey therefore asked (1) whether funding agencies should require grantees to attest on grant applications that they have considered the dual use implications of their proposed research and (2) whether they believed funding agencies would be less likely to fund grant proposals if the proposed research has dual use potential.

Discussion

Figure 3-9 presents data from respondents regarding their views about the role of funding agencies.

Almost 60 percent of the life scientists who responded to the survey agreed or strongly agreed that funding agencies should ask researchers to attest that they had considered the dual use potential of their work. These life scientists may view treating dual use research issues in grant applications as similar to existing procedures for human, recombinant DNA, and animal research. All of these issues require a paragraph demonstrating that PIs have considered the possible implications of their research and alternative strategies to alleviate any concerns. Examples of comments from respondents are shown in Box 3-9.

The question about whether disclosing the dual use potential of one's research would negatively affect funding decisions did not elicit consensus as noted in Figure 3-10. Over 40 percent of respondents disagreed or strongly disagreed that identifying the dual use potential of one's research

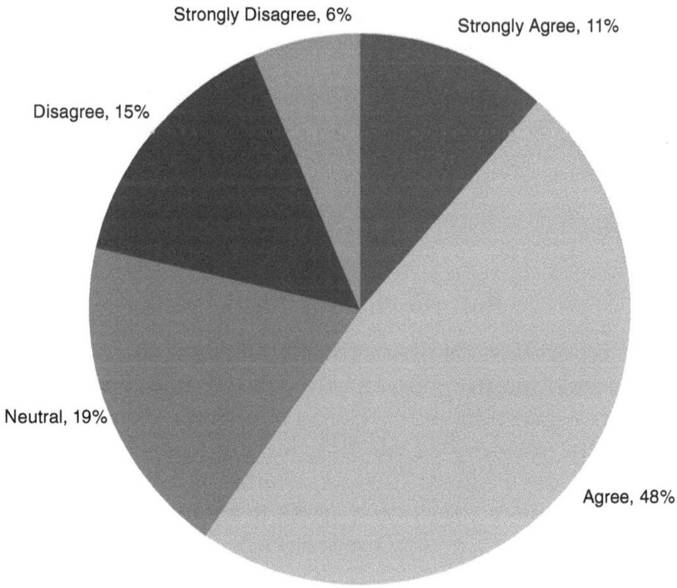

FIGURE 3-9 Respondents' views on whether funding agencies should require grantees to attest on grant applications that they have considered dual use implications of their proposed research.
NOTE: Based on 1,633 respondents.
SOURCE: NRC/AAAS Survey; data tabulations by staff.

BOX 3-9
Illustrative Respondent Comments on Review of Grant Applications

"I believe strongly in freedom to do research but also believe the safety of the US and world trumps this freedom. To be effective, restrictions should avoid busy work and apply only to directly applicable research. Funding agencies are in the best position to flag applications for dual use research."

"Grants should be reviewed by those knowledgeable in dual use experiments of concern and this information communicated to PIs for consideration and dissemination to PI's home institution and employees."

"Dual research should be submitted and reviewed in special study sections and monitored closely esp. as to research personnel involved."

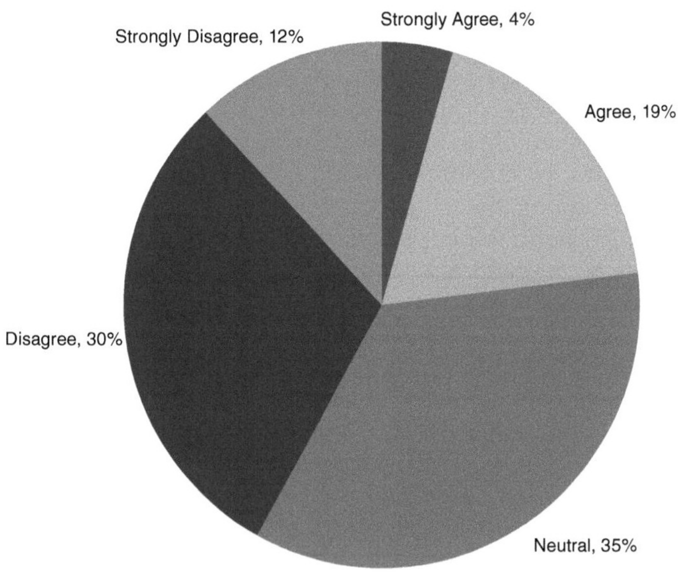

FIGURE 3-10 Respondents' views on whether funding agencies would be less likely to fund grant proposals if the proposed research has dual use potential.
NOTE: Based on 1,633 respondents.
SOURCE: NRC/AAAS Survey; data tabulations by staff.

TABLE 3-11 Correlation Between Respondents' Views on the Role of Specific Institutions and of Other Organizations' Responsibility

Variable	Variable: Institutions Should Provide Training	Grant Proposals Should Be Reviewed	Funding Agencies Should Require Attestation	Professional Societies Should Have Codes of Conduct
Institutions should provide training	1			
Grant proposals should be reviewed	0.4 (1,633)	1		
Funding agencies should require attestation	0.44 (1,633)	0.47 (1,633)	1	
Professional societies should have codes of conduct	0.19 (1,503)	0.17 (1,499)	0.27 (1,499)	1

NOTE: Number of respondents in parentheses.
SOURCE: NRC/AAAS Survey; data tabulations by staff.

would have a negative impact on the likelihood that grant proposals would be funded. About a quarter of the respondents (23 percent) agreed or strongly agreed that funding agencies would be less likely to fund research proposals that acknowledged dual use potential, and 35 percent were neutral about the potential impact on funding decisions.

A correlation was calculated to find out whether respondents who supported a requirement to attest to dual use potential on grant proposals also expect that such disclosure would *not* have a negative impact on the prospects for funding. In fact, no relationship was found (Spearman's $\rho = -0.02$, $n = 1{,}633$). Support or opposition for disclosing dual use potential in a grant application seems to be unrelated to whether they think it would affect awarding of the grants.

The committee also wondered if respondents who thought that one group (professional societies, one's home institution, or funding agencies) should take some responsibility for addressing dual use issues thought the other groups should as well. Four questions were compared: support for professional society codes of conduct; home institution review of grant proposals; home institution training; and funding agency requirements for applicant attestation. Spearman's ρ was used and the results are shown in Table 3-11. There are positive correlations between the four pairs of variables. In general, those who support one type of institutional role also tend to support the others.

Summary of Key Results

- About 60 percent of the respondents agreed or strongly agreed that funding agencies should ask researchers to attest that they had considered the dual use potential of their work (Figure 3-9).
- Forty-two percent of the respondents did not think that funding agencies would be less likely to fund grant proposals where the research had dual use potential; only 23 percent felt that there would be a negative impact on funding research—the remainder were uncertain (Figure 3-10).
- Respondents who supported a role for funding agencies in dual use research oversight also tended to support a role for other organizations (i.e., home institutions and professional societies) (Table 3-11).

Role for the Federal Government

As discussed in some detail in Chapter 1, one of the most contentious aspects of the proposals for oversight of dual use research is the role of the federal government. A number of models have been offered by which oversight might be carried out, with mixes of voluntary, regulatory, and statutory provisions.

Discussion

The survey asked whether more federal oversight of research was needed. As shown in Figure 3-11, the respondents did not support an increase in oversight by the federal government; this may be because some equated federal oversight with mandatory regulation, such as the select agent and export control regulations. Thirty-five percent disagreed and 11 percent strongly disagreed compared to 22 percent who agreed and only 4 percent who strongly agreed with increased federal oversight. It may be worth noting that 27 percent reported that they were neutral, which might suggest opportunities for discussion and debate among those not already committed. See Box 3-10.

Correlation analyses were conducted to test for an association between support for greater federal oversight, individual responsibility, and roles for institutions and organizations. Note that these analyses test attitudes regarding increased federal oversight versus other types of responsibilities. They do not test individual versus institutional responsibilities. The results are shown in Table 3-12. The table indicates that respondents who saw a need for greater federal oversight also tended to agree that others (both scientists and institutions) should play a role but supporting an individual and institutional role in oversight of dual use research does not necessarily imply a role for the federal government.

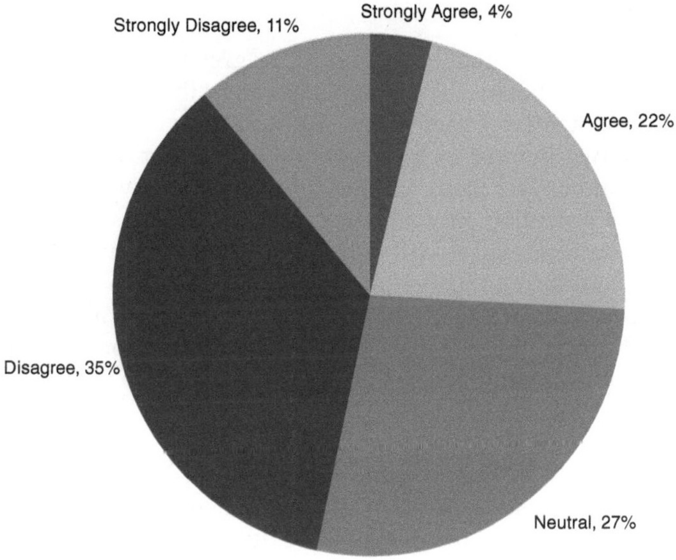

FIGURE 3-11 Respondents' views regarding whether dual use research needs greater federal oversight.
NOTE: Based on 1,637 respondents.
SOURCE: NRC/AAAS Survey; data tabulations by staff.

BOX 3-10
Illustrative Respondent Comments on Federal Oversight

"Dual use is nothing new, and restrictions on research in the name of preventing a bioterrorist attack is the worst form of censorship and is far more likely to produce greater problems and retard research than it is to foil a potential terrorist—free societies need to remain free, and research needs to proceed without additional controls beyond what is needed to ensure the safety of those performing the research and the general public from the research itself—not unlikely hypothetical misuse of the research by malignant boogie men. By such reasoning, all research could potentially be censored and we could enter a new dark ages."

"Any desire to restrict the scientific pursuit of items that can be ambiguously termed "dual-use" would be harmful to science in this country. Additional rules for science usage are much riper for abuse of scientists than the "dual-use" science in the public domain."

"By restricting research on biological agents that could be used in bioterrorism, we are more likely to prevent knowledge that will protect us from such agents."

TABLE 3-12 Correlation Between Respondents' Views Regarding Need for Greater Federal Oversight and Other Actors' Responsibility

Correlation Between Need for Greater Federal Oversight and . . .	Spearman's ρ	*N*
PIs should be responsible for initial evaluation of their research	0.03	1,637
Scientists should assure their institutions	0.30	1,637
Scientists should take an oath	0.35	1,637
PIs should train their staff, students, etc.	0.09	1,637
Institutions should provide mandatory training	0.36	1,637
Institutions should review grant proposals	0.38	1,633
Funding agencies should require attestation	0.38	1,633

SOURCE: NRC/AAAS Survey; data tabulations by staff.

Different types of respondents might show different levels of support for greater federal oversight. Two methods for disaggregating respondents were used: first, the type of research they engaged in; and second, where they were employed. The results are summarized in Table 3-13; the complete results can be found in Appendix Table D-3.

As the table shows, respondents who considered their research to be dual use or to involve one of the seven categories of experiments, or who work with select agents were slightly more likely to disagree that greater federal oversight of research was needed. There appeared to be no difference among respondents on the basis of where they worked.

In a separate survey addressing a somewhat different question, Professor Victoria Sutton of the Texas Tech University School of Law conducted a national in 2008 survey of PIs and sub-PIs in NIH regional centers of excellence. Professor Sutton asked the researchers their opinions concerning the select agent rules (42 CFR §73), and the effectiveness of these rules in achieving their regulatory goal of national security and protecting public health. Of the 201 out of 509 who responded, 93.4 percent agreed that select agents should be regulated (Sutton 2009). These findings support the conclusion Sutton drew from an earlier regional study that found similar levels of agreement: "This tends to disprove some perceptions among policymakers that biodefense researchers oppose the regulations simply because they do not want to be regulated" (Sutton 2007).[27]

[27]Note: The support for regulation of select agents does not reflect satisfaction with the current implementation of the select agent regulations, which Sutton and others have reported are burdensome (Council on Government Relations 2003) and problematic for the life sciences community (see Fischer 2006 for a full discussion of the impact of the current select agent regulations on universities and the life sciences community).

TABLE 3-13 Average Rating, on a Scale of 1 to 5, of Respondents' Agreement with Statement That Greater Federal Oversight Is Needed, by Type of Research and Employment

Variable	Average
Works with dual use?	
Yes	2.5
No	2.7
Works with seven categories of experiments?	
Yes	2.5
No	2.7
Works with select agents?	
Yes	2.6
No	2.8
Employer type	
Industry	2.8
Academia	2.7
Government	2.8
Other	2.8

NOTE: On a scale from 1 (Strongly Disagree) to 5 (Strongly Agree).
SOURCE: NRC/AAAS Survey; data tabulations by staff.

Summary of Key Results

- Only 26 percent of life scientists who responded to the survey felt that dual use research needs greater federal oversight (Figure 3-11).
- Life scientists who responded to the survey did not see responsibility as an "either-or" proposition: Those that supported greater responsibility by other institutions also tended to support greater federal oversight. For example, those who agreed with a need for greater federal oversight also tended to agree that institutions should review grant proposals (Table 3-12).
- Life scientists who responded to the survey who indicated that their research was dual use or that they were working with, or had worked with, select agents were slightly less likely to agree that greater federal oversight is needed.

POLICY

The final set of questions focused on what policy steps scientists would support to reduce the potential that knowledge, tools, or techniques from dual use research could pose a threat to national security. The survey participants were asked whether they would support seven possible measures:

• Certification of researchers conducting dual use research;
• Greater restrictions on access to specific biological agents or toxins;
• Licensure of certain biological equipment that is commonly used in life sciences research;
• Restrictions on disclosure of details about the research or its findings through personal communication;
• Alteration or removal of certain experimental methods or findings prior to publication or presentation;
• Restrictions on publication of findings based on dual use potential; and
• Classification of research findings based on dual use potential.

Discussion

The respondents' views are depicted in Figure 3-12, again displayed so that those measures with the most "Strongly Agree" responses appear

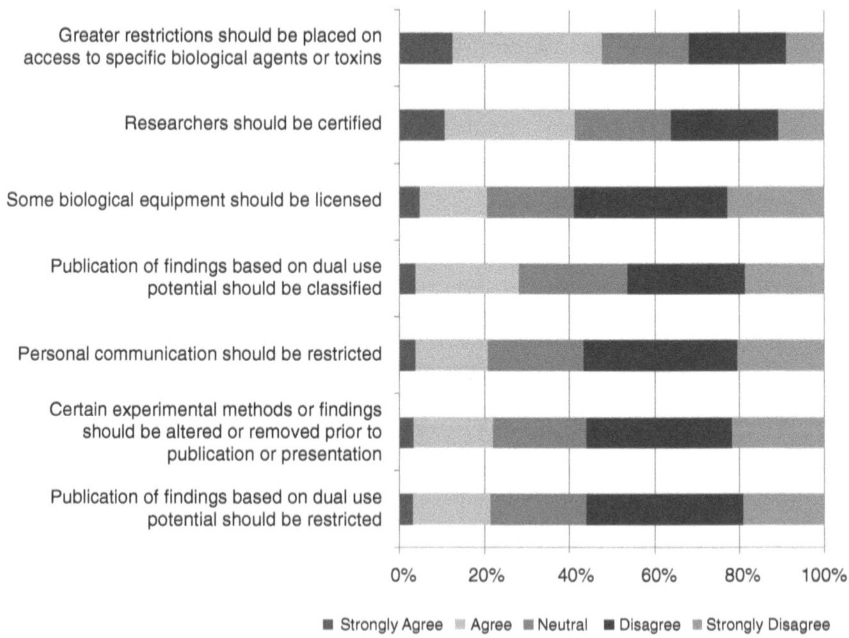

FIGURE 3-12 Respondents' views on steps that should be taken to prevent the potential that knowledge, tools, or techniques from dual use research could pose a threat to national security.
NOTE: Based on 1,658 respondents.
SOURCE: NRC/AAAS Survey; data tabulations by staff.

at the top. There was no clear consensus about proposed measures; none of the proposed measures was able to attract the support of a majority of respondents, although increasing restrictions on access to pathogens received almost 50 percent support. Some illustrative comments from respondents are provided in Box 3-11.

Respondents might have different views regarding whether certain policies should be required based upon where they were employed or what type(s) of research they were performing. Two methods for disaggregating respondents were used: first, the type of research they engaged in; and second, where they were employed. The results are summarized in Table 3-14; complete data can be found in Appendix Table D-4.

As the table shows, respondents who conduct one of the three types of research appeared slightly more likely to disagree that many of the

BOX 3-11
Illustrative Respondent Comments on Policy Measures

"The federal government should monitor the potential threat of dual purpose results but should not interfere with the scientific process of publication and research."

"It's hard enough to do research, and additional controls based on dual use fear-mongering will make it even harder."

"We should remember that several outbreaks of SARS and foot-and-mouth disease are from labs working on these agents. It will not be surprising at all that acts of bioterrorism may eventually be committed by members of research labs where these agents are being studied. The likelihood of state-sponsored bioterrorism, in my opinion, is low. In this sense, persons who conduct research of dual use biotechnology should be subjected to security clearance to safe-guard the appropriate use of the technology."

"Bioterrorism is a real and horrible threat to all of us. We all want to be safe. The challenge for all of us is to implement measures that will be meaningful and effective—not "feel good" approaches that will inhibit research without making a positive impact on safety."

"The risks are real but I worry that the "solution" could be worse."

"If I had one sentence, it would include a caution that how we anticipate and prevent such a threat from occurring is being driven by reasoning rather than by fear."

TABLE 3-14 Average Rating, on a Scale of 1 to 5, of Respondents' Agreement with Statement That a Particular Policy Is Needed, by Type of Policy and Respondent

Variable	Certification of Researchers	Restrictions on Access	Licensure of Equipment	Restrictions on Personal Communication	Modification of Manuscripts or Presentations	Restrictions on Publications	Classification of Findings
Works with dual use?							
Yes	2.8	2.9	2.2	2.2	2.4	2.3	2.3
No	3.1	3.2	2.4	2.4	2.4	2.5	2.6
Works with seven categories of experiments?							
Yes	2.8	2.9	2.1	2.5	2.5	2.2	2.4
No	3.0	3.1	2.4	2.4	2.4	2.4	2.6
Works with select agents?							
Yes	3.0	3.0	2.3	2.4	2.5	2.5	2.6
No	3.1	3.3	2.5	2.5	2.5	2.5	2.7
Employer type							
Industry	3.0	3.4	2.5	2.7	2.7	2.7	2.9
Academia	3.0	3.1	2.3	2.4	2.4	2.4	2.6
Government	3.1	3.4	2.6	2.6	2.6	2.5	2.7
Other	3.0	3.2	2.6	2.4	2.5	2.4	2.8

NOTE: On a scale from 1 (Strongly Disagree) to 5 (Strongly Agree).
SOURCE: NRC/AAAS Survey; data analysis by staff.

policies should be required. No difference was discerned regarding where the respondents were employed.

The committee wondered if there might be an association between scientists' responses to this question and their views about the role of different types of information in allowing individuals to create dangerous biological agents. We hypothesized that relationships might exist between

1. Scientists' support for altering or removing methods or findings prior to publication or presentation and positive answers to presentations providing sufficient information,
2. Scientists' support for restrictions on publications and positive answers to scientific journal articles providing sufficient information,
3. Scientists' support for restrictions of disclosure of details through personal communications and positive answers to personal communications providing sufficient information.

The results of a correlation analysis using Spearman's ρ to test these relationships are provided in Table 3-15. As Table 3-15 shows, there is only a small positive relationship between supporting a restriction on a source of information and believing that the source is particularly useful to people with malevolent intentions. It is worth remembering that none of the policy options here, as opposed to some of the other measures discussed in earlier sections, attracted strong support, so it could be the

TABLE 3-15 Correlation Between Respondents' Views on Policy Options and Their Views About the Role of Different Types of Information in Allowing Individuals to Create Dangerous Biological Agents

	Source of Information		
Support for Restriction	Journals	Presentations	Personal Communication
Restrictions on disclosure of details through personal communication			0.18 (1,107)
Alteration or removal of methods or findings prior to publication or presentation	0.12 (1,266)	0.11 (1,294)	
Restrictions on publication of findings	0.16 (1,266)		

NOTE: The number of respondents is listed in parentheses.
SOURCE: NRC/AAAS Survey; data analysis by staff.

scientists' doubts about the measures themselves that explain the weak relationship.

Summary of Key Results

- None of the seven policy measures attracted a majority of support from survey respondents, although increasing restrictions on access to pathogens received almost 50 percent support (Figure 3-12).
- Respondents who worked with dual use research often were less likely to favor the policies (Table 3-14).

CONCLUDING REMARKS

This chapter has presented the results of our survey of a sample of AAAS members in the life sciences. The key results for each topic addressed in the survey have been presented. In a number of cases, additional statistical analyses were performed to explore possible explanations for the results and these are presented in the relevant sections. Some of the results support assumptions that the committee made about likely responses and some of the results were surprising to at least some of the committee members. The next chapter presents a synthesis of the key results from the survey as well as the committee's conclusions about what they may mean for policies to reduce the risks that the results of research in the life sciences will be used for malign purposes. The chapter concludes with the committee's recommendations for further research and actions related to outreach and education.

4

Conclusions and Recommendations

INTRODUCTION

The results of the survey provide some of the first empirical data about the perceptions of a sample of U.S. life scientists of the potential risks of misuse of legitimate scientific research for malicious purposes. The survey obtained information from a diverse group of academic, government, and industry researchers. The survey data provide evidence about how the respondents perceive the sources of risk related to dual use research, the actions that some scientists are taking to reduce the risk of misuse of science, and the prospects for acceptance of various policy proposals aimed at reducing the risks of misuse of legitimate life science research.

While useful, the results of the survey must be viewed with caution because of the low response rate and possible response bias. Scientists who may be involved in biodefense or select agent research, for example, may be more aware of the dual use dilemma and thus more likely to have responded to the survey. In addition, a few of the questions could have been interpreted in multiple ways. Despite the limitations, which are discussed in detail in Chapter 2, the committee believes that the data obtained in this study offer valuable insights and new information.

Overall, the survey findings suggest that there is considerable support for models of oversight that rely on the responsible conduct of research and self-governance by the scientific community. The responses also suggest, however, that there is a critical need to clarify the scope of research activities of high concern and to determine the appropriate actions that

members of the life sciences community can take to reduce the risk of misuse of science for bioweapons development or bioterrorism.

The rest of the chapter provides a summary of the survey findings. Following a brief summary of the perceptions of risks of the scientists who responded to the survey, three key areas of current and potential activities and policies are highlighted: actions that life scientists have already taken to address dual use concerns, mechanisms for the oversight of research, and issues related to education and outreach. The chapter closes with the committee's recommendations for furthering education and outreach activities that are based on the findings of the survey and its own judgments and analysis.

PERCEPTIONS OF RISK

The findings suggest that, on average, the scientists who responded to the survey perceive a potential, but not overwhelming, risk of bioterrorism and that the risk is greater outside the United States. On average, the respondents believed that there is a 51 percent chance that there will be an act of bioterrorism somewhere in the world in the next 5 years and a 35 percent chance that there will be an act of bioterrorism in the United States in the next 5 years. Three-quarters of the respondents believe that a preference for other means of attack is the primary reason why there have been only a few acts of bioterrorism to date; overwhelmingly, 87 percent of respondents said that they believe that terrorists are not deterred by the threat of being caught and punished. Fewer scientists considered a lack of knowledge (46 percent) or access to equipment (51 percent) or agents (36 percent) to be significant barriers. It may be that one's perceived risk of such an attack is related to one's support for taking measures to reduce the risks that life sciences research might be misused.

With regard to the chance that the knowledge, tools, or techniques from dual use research will facilitate bioterrorism, the respondents perceive a 28 percent chance, on average, of such a bioterror attack within the next 5 years. Half of the respondents thought that if someone wanted to create a harmful biological agent, the Internet would be the most likely place to provide sufficient information for life scientists with college-level training. Other sources of information—articles in scientific journals (40 percent), personal communications (38 percent), and presentations at professional meetings (18 percent)—were considered relatively less likely sources, although on average 45 percent of respondents answered "Don't Know" to these questions.

ACTIONS TAKEN BY LIFE SCIENTISTS IN RESPONSE TO DUAL USE CONCERNS

Although the responses to the survey indicate that bioterrorism probably is not perceived to present a serious immediate risk to U.S. or global security, the survey results also indicate that there is already concern about dual use issues among some of the life scientists who responded. Fifteen percent of the respondents (260 individuals out of 1,744) indicated that they are so concerned about dual use research that they have taken actions, even in the absence of guidelines or mandatory regulations from the U.S. government. Some respondents reported that they had broken collaborations, not conducted some research projects, or not communicated research results. The results indicate that more scientists have modified their research activities than some members of the committee expected on the basis of previous reports of manuscripts that have been modified or not published because of dual use concerns.

Interestingly, many of the actions that the respondents reported taking to mitigate concerns occurred before the publication stage; much of the behavior change occurred during the research design, collaboration, and early communication stages. Of particular interest and concern to the committee, a few respondents commented on their concerns about foreigners as potential security risks, which may be reflected in the reported avoidance of some collaborations.

The survey results suggest that: (1) some life scientists in the United States may be willing to consider self-governance aimed at the responsible scientific conduct for dual use research, and (2) some life scientists in the United States are already acting, even in the absence of government regulations and guidance, to protect against the perceived risk of misuse of dual use research.

OVERSIGHT MECHANISMS

With a proposed oversight framework for dual use research of concern proposed by NSABB in June 2007 now under consideration within the U.S. government, the survey was an opportunity to assess scientists' attitudes toward specific policy options. Many of the respondents indicated that they believe that personal responsibility, including measures such as codes of conduct, could foster a positive culture within the scientific community to evaluate the potential consequences of their research for public safety and national security. They also indicated that they believe that individual researchers, professional scientific societies, institutions, and scientific journals should be responsible for evaluating dual use potential of research and/or fostering the culture of scientific responsibility.

A majority of those who responded to the survey favored self-gov-

ernance mechanisms for dealing with dual use research of concern, such as those proposed by the Fink report (NRC 2004a), rather than additional mandatory government regulations. In addition to the low level of support for greater federal oversight (26 percent), the individual comments indicated a belief that increased government oversight of dual use research would be counterproductive by inhibiting the research needed to combat emerging infectious diseases and bioterrorism as well as being potentially harmful to the scientific enterprise more generally.

The survey suggests that most of the respondents (82 percent) favor their professional societies' prescribing a code of responsible conduct to help prevent misuse of life sciences research. However, many respondents (66 percent) did not know whether the societies to which they belonged already had codes that address dual use issues, and some of the societies most frequently cited do not in fact have a code. There was substantially less support (38 percent agree or strongly agree) for a Hippocratic-style oath.

The results also indicate potential support for journals having biosecurity policies. Yet, most of the respondents did not know if any of the journals in which they have published or to which they have submitted manuscripts have those policies. Moreover, more than half of those who responded to the survey strongly disagreed or disagreed with restrictions on personal communication, altering or removing methods or findings from scientific publications, or limiting publication itself.

The survey points to a likely preference for self-governance measures to provide oversight of dual use research. There was substantially less support for mandatory measures that might be imposed by regulation, although the results varied for different policy measures. The results indicate that there may be greater support for restrictions on access to biological agents (just under 50 percent of the respondents said they agree or strongly agree) and certifications of researchers (just over 40 percent of the respondents said they agree or strongly agree) than for any control of scientific knowledge generated from the research or through information exchange (only 20 to 30 percent of respondents supported these measures). Table 4-1 provides a list of the level of support for the various measures addressed in the survey.

The survey results suggest there is *support* for:

1. Greater oversight that is not federally mandated,
2. Self-governance mechanisms as an approach for preventing misuse of life science research and knowledge,
3. Professional and scientific societies adopting codes of conduct that include dual use research as suggested in the Fink report (NRC 2004a),
4. Establishing and implementing policies for authors and reviewers

TABLE 4-1 Summary of Results Regarding Support for Measures of Personal and Institutional Responsibility

Measures of Personal or Institutional Responsibility	Strongly Agree or Agree (or Respond Yes*) (%)
Principal investigators should be responsible for the initial evaluation of the dual use potential of their life sciences research.	87
Principal investigators should be responsible for training lab staff, students, and visiting scientists about dual use research.	86
Should professional science societies have codes for the responsible conduct of dual use life sciences research?	82*
University and college students should receive educational lectures and materials on dual use life sciences research.	68
Scientists should provide formal assurance to their institution that they are assessing their work for dual use potential.	67
Funding agencies should require grantees to attest on grant applications that they have considered dual use implications of their proposed research.	60
Should scientific journals have policies regarding publication of dual use research?	57*
Institutions should provide mandatory training for scientists regarding dual use life sciences research.	55
Greater restrictions should be placed on access to specific biological agents or toxins.	47
Researchers conducting dual use research should be certified.	42
All grant proposals for life sciences research with dual use potential should be reviewed by a researcher's institution prior to submission for funding.	41
Scientists conducting or managing research should take an oath.	38
Research findings should be classified based on their dual use potential.	28
Dual use research needs greater federal oversight.	26
Certain experimental methods or findings should be altered or removed prior to publication or presentation.	22
Certain biological equipment that is commonly used in life science research should be licensed.	21
There should be restrictions on disclosure of details about the research or its findings through personal communication.	21
There should be restrictions on publication of findings based on their dual use potential.	21

SOURCE: NRC/AAAS Survey of Life Scientists; data analysis by staff.

to consider the dual use potential of research manuscripts submitted to journals.

The survey results suggest there is *opposition* to:

1. Mandatory government regulations to govern the conduct of dual use research and the communication of knowledge from that research;
2. Other mandatory oversight actions, such as oaths or licensing of scientists.

Based on the survey results and its own analysis, the committee believes that a basis of support exists within the U.S. scientific community for measures that, taken together, could lead to the development of a system of self-governance for the oversight of key aspects of dual use research.

EDUCATION AND OUTREACH

A major reason for conducting the survey was to inform efforts for education and awareness-raising about dual use research by providing empirical data on the attitudes of a sample of the life sciences community. In general, the respondents to this survey would likely support educational and outreach activities aimed at raising awareness of the dual use dilemma. The respondents indicated that they supported educational materials and lectures on dual use research for students. They also supported mandatory training by institutions for practicing life scientists regarding dual use research of concern.

The survey results also highlight the need to better define the scope of dual use research of concern. Fewer than half of the respondents who indicated that they were carrying out dual use research activities felt that their research fell into one of the seven categories of research of concern specified by the NSABB. The dual use experiments of concern as listed in the Fink report (NRC 2004a) and by the NSABB are all based on microbial research, but other relevant research, such as theoretical research, scenario development, or applied research (e.g., pharmaceutical formulations or neuroscience research) can be of dual use concern. In their individual comments, a number of respondents stressed the difficulties of defining dual use, as did participants in the focus groups used to develop the survey. Clearly a better understanding of the scope of dual use research of real concern would help any educational or outreach activities aimed at raising the awareness of life scientists so that appropriate actions can be taken.

Based on the survey results and its own analysis, the committee believes that there is support for mandatory education and training about dual use issues, most likely as part of ethics and responsible conduct of research training.

RECOMMENDATIONS

The committee believes that the survey raises several hypotheses that merit further research about the views of life scientists about oversight policies and education and outreach efforts to address concerns about dual use issues in the life sciences. In particular, based on the survey results and its own deliberations, the committee offers the following recommendations:

Oversight, Education, and Outreach

1. Explore how to continue and to expand the dialogue within the life sciences community about dual use research of concern.
2. Explore ways to provide guidance to the life sciences community about appropriate actions that can be taken to protect against the misuse of dual use research.
3. Seek to better define the scope of knowledge in the life sciences that may be at greatest risk for misuse and to provide the life sciences community with criteria for recognizing dual use research of concern.
4. Encourage journals that have biosecurity policies or plan to adopt them in the future and the professional and scientific societies that have or plan to develop codes of conduct to communicate those policies more effectively.

Further Research

1. Examine the effectiveness of existing educational programs and how they can be enhanced and focused.
2. Seek to extend educational and awareness-raising efforts being conducted in the United States to the broad international scientific community.
3. Examine how education and outreach activities can be developed to guide the life science community's response to concerns about dual use research so as to ensure that actions taken by the community are appropriate and contribute to advancing scientific knowledge while protecting national security.
4. Conduct additional surveys, interviews, or focus groups of U.S. life scientists that better represent the full community, with higher response

rates than the current study was able to achieve, and the ability to assess potential bias, in order to gain

i. a better understanding of the awareness of a broader range of U.S. life scientists about dual use research of concern and the measure that they would support to reduce the threat that research in the life sciences could be subverted to do harm;

ii. a better understanding of the types of behavioral changes being made in response to dual use concerns to determine if actions by life scientists are contributing to national security or harming scientific research; such research is critical given the actions that the current survey suggests are being taken;

iii. more detailed information about the types of changes scientists are making or scientists' thoughts about dual use issues, experiments of concern, and select agents;

iv. a better understanding of scientists' experiences with education on this topic and their views about the content and delivery of educational and training materials.

5. Conduct additional surveys of life scientists outside the United States that would enable comparisons of attitudes toward dual use research of concern and inform educational and outreach programs so that they can be effective on a global scale. Such knowledge could also facilitate international discussions of potential measures to address dual use concerns.

Bibliography

AAPOR (American Association for Public Opinion Research). 2006. *Standard Definitions: Final Dispositions of Case Codes and Outcome Rates for Surveys*, 4th ed. Lenexa, KS: AAPOR.

Alberts, B. 2005. Modeling attacks on the food supply. *Proceedings of the National Academy of Sciences* 102:9737–9738.

Alberts, B., W. A. Wulf, and H. Fineberg. 2002. Current visa restrictions interfere with U.S. science and engineering contributions to important national needs. Statement from the presidents of the National Academies. December 13; rev. June 13, 2003. Available at http://www8.nationalacademies.org/onpinews/newsitem.aspx?RecordID=s12132002.

Alibek, K., and S. Handelman. 1999. *Biohazard: The Chilling True Story of the Largest Covert Biological Weapons Program in the World—Told from Inside by the Man Who Ran It*. New York: Dell.

Annan, K. 2006. Uniting Against Terrorism: Recommendations for a Global Counter-Terrorism Strategy. Report of the Secretary-General. A/60/825. New York: United Nations. Available at http://www.un.org/unitingagainstterrorism/contents.htm.

Annas, G. J. 2006. Bioterror and "bioart"—A plague o' both your houses. ***New England Journal of Medicine*** 354:2715–2720.

Atlas, R. 2003. Preserving scientific integrity and safeguarding our citizens: Challenges for scientific publishers in the age of bioterrorism. Available at http://www7.nationalacademies.org/dsc/Atlas_Presentation_Sci_Openness.pdf.

Atlas, R., and M. Dando. 2006. The dual use dilemma for the life sciences: Perspectives, conundrums, and global solutions. *Biosecurity and Bioterrorism: Biodefense Strategy, Practice, and Science* 4:276–286.

BBSRC/MRC/Wellcome Trust (Biotechnology and Biological Sciences Research Council, the Medical Research Council, and the Wellcome Trust). 2005. Managing Risks of Misuse Associated with Grant Funding Activities. September. Available at http://www.bbsrc.ac.uk/organisation/policies/position/public_interest/misuse_of_research_joint.pdf.

Berk, R. A., S. G. Korenman, and N. S. Wenger. 2000. Measuring consensus about scientific research norms. *Science and Engineering Ethics* 6:315–340.

Bernauer, H., J. Christopher, W. Deininger, M. Fischer, P. Habermeier, K. Heumann, S. Maurer, S. Schwer, P. Stähler, and T. Wagner. 2008. Technical Solutions for Biosecurity in Synthetic Biology. Report of a Workshop. Munich, Germany. Available at http://www.ia-sb.eu/wp-content/uploads/2008/09/iasb_report_biosecurity_syntheticbiology.pdf.

Block, S. M. 1999. Living nightmares: Biological threats enabled by molecular biology. Pp. 39–75 in *The New Terror: Facing the Threat of Biological and Chemical Weapons*, S. Drell, A. D. Sofaer, and D. Wilson, eds. Stanford, CA: Hoover Institution Press.

BLS (Bureau of Labor Statistics). 2008. The 2006–16 job outlook in brief. *Occupational Outlook Quarterly* 52(Spring):1.

Blumenthal, D., E. G. Campbell, N. Causino, and K. S. Louis. 1996a. Participation of life-science faculty in research relationships with industry. *New England Journal of Medicine* 335:1734–1739.

Blumenthal, D., N. Causino, E. Campbell, and K. S. Louis.1996b. Relationships between academic institutions and industry in the life sciences—An industry survey. *New England Journal of Medicine* 334:368–373.

Blumenthal, D., E. G. Campbell, M. S. Anderson, N. Causino, K. S. Louis. 1997. Withholding research results in academic life science. Evidence from a national survey of faculty. *Journal of the American Medical Association* 277:1224–1228.

Blumenthal, D., E. G. Campbell, M. Gokhale, R. Yucel, B. Clarridge, S. Hilgartner, N. A. Holtzman. 2006. Data withholding in genetics and the other life sciences: Prevalences and predictors. *Academic Medicine* 81(2):137–145.

Borras, C. 1982: AAAS Council Meeting. *Science* 215:1069–1072.

Borras, C. 1984: AAAS Council Meeting. *Science* 225:387–390.

Bügl, H., J. P. Danner, R. J. Molinari, J. Mulligan, D. A. Roth, R. Wagner, B. Budowle, R. M. Scripp, J. A. L. Smith, S. J. Steele, G. Church, and D. Endy. 2006. A practical perspective on DNA synthesis and biological security. Presentation at the International Consortium for Polynucleotide Synthesis. December 4. Available at http://pgen.us/ICPS.htm.

Bullock, M., and S. Panicker. 2003. Ethics for all: Differences across scientific society codes. *Science and Engineering Ethics* 9:159–170.

Campbell, P. 2006. Empowerment and restraint in scientific communication. *EMBO Reports* 7(Special Issue):S18–S22.

Campbell, E. G., B. R. Clarridge, M. Gokhale, L. Birenbaum, S. Hilgartner, N. A. Holtzman, and D. Blumenthal. 2002. Data withholding in academic genetics: Evidence from a national survey. *Journal of the American Medical Association* 287:473–480.

Carus, W. S. 2000. The Rajneeshees (1984). Pp. 115–138 in *Toxic Terror: Assessing Terrorist Use of Chemical and Biological Weapons*, J. Tucker, ed. Cambridge, MA: MIT Press.

CDC (Centers for Disease Control and Prevention). 2005. Researchers reconstruct 1918 pandemic influenza virus; effort designed to advance preparedness. Press release, October 5. Available at http://www.cdc.gov/od/oc/media/pressrel/r051005.htm.

Cello, J., A. V. Paul, and E. Wimmer. 2002. Chemical synthesis of poliovirus cDNA: generation of infectious virus in the absence of natural template. *Science* 297:1016–1018.

Check, E. 2002a. Biologists apprehensive over U.S. moves to censor information flow. *Nature* 415(6874):821.

Check, E. 2002b. U.S. prepares ground for security clampdown. *Nature* 418(6901):906.

Cohen, J. 2002. Designer bugs. *Atlantic Monthly* (July/August). Available at http://www.theatlantic.com/issues/2002/07/cohen-j.htm.

Commission on the Prevention of WMD Proliferation and Terrorism. 2008. World at Risk: Report of the Commission on the Prevention of WMD Proliferation and Terrorism, Authorized Edition. New York: Vintage Books. Available at http://www.preventwmd.gov/report/.

Cook, C., F. Heath, and R. Thompson. 2000. A Meta-analysis of Response Rates in Web- or Internet-based Surveys. *Educational & Psychological Measurement*, 60(6):821–837.

Corneliussen, F. 2005. The impact of regulations on firms: A study of the biotech industry. *Law & Policy* 27:429–449.

Corneliussen, F. 2006. Adequate regulation, a stop-gap measure, or part of a package? *EMBO Reports* 7(Special Issue):S50–S54.

Council for Excellence in Government. 2004. From the Home Front to the Front Lines: America Speaks Out About Homeland Security Survey. March. Available at http://www.excelgov.org/admin/FormManager/filesuploading/Homeland_Full_Report.pdf.

Council on Governmental Relations. 2003. Report of the Working Group on the Cost of Doing Business. Available at http://rbm.nih.gov/fed_reg_20030906/FRNotice/Associations/AAU_A-21_CODB.pdf.

CSIS (Center for Strategic and International Studies). 2006. *The Biological Weapons Threat and Nonproliferation Options: A Survey of Senior U.S. Decision Makers and Policy Shapers.* Washington, DC: CSIS and the Carnegie Endowment for International Peace.

Curtin, R., S. Presser, and E. Singer. 2000. The effects of response rate changes on the index of consumer sentiment. *Public Opinion Quarterly* 64:413–428.

Dam, K., and H. Lin (eds.). 1996. *Cryptography's Role in Securing the Information Society* [CRISIS]. Washington, DC: The National Academies Press.

Dando, M. R., and B. Rappert. 2005. Codes of Conduct for the Life Sciences: Some Insights from UK Academia. Bradford Briefing Paper 16. Available at http://www.brad.ac.uk/acad/sbtwc/briefing/BP_16_2ndseries.pdf.

Dando, M. R., and J. Revill. 2005. Raising Awareness: A Hippocratic Oath for Life Scientists. Bradford Briefing Paper 18. Available at http://www.brad.ac.uk/acad/sbtwc/briefing/BP_18_2ndseries.pdf.

Davidson, E. M., R. Frothingham, and R. Cook-Deegan. 2007. Practical experiences in dual use review. *Science* 216:1432–1433.

De Vries, R., M. S. Anderson, and B. C. Martinson. 2006. Normal misbehavior: Scientists talk about the ethics of research. *Journal of Empirical Research on Human Research Ethics* 1(1):43–50.

Diffie, W. 1996. Foreword. In *Applied Cryptography: Protocols, Algorithms, and Source Code* in C, 2nd ed., B. Schneier, ed. Hoboken, NJ: John Wiley and Sons. Available at http://www.schneier.com/book-applied-foreword.html.

Dillman, D. A. 1991. The design and administration of mail surveys. *Annual Review of Sociology* 17:225–249.

Eastwood, S., P. Derish, E. Leash, and S. Ordway. 1996. Ethical issues in biomedical research: Perceptions and practices of postdoctoral research fellows responding to a survey. *Science and Engineering Ethics* 2(1):89–114.

Epstein, G. L. 2001. Controlling biological warfare threats: Resolving potential tensions among the research community, industry, and the national security community. *Critical Reviews in Microbiology* 27:321–354.

FBI (Federal Bureau of Investigation). 2008. Science briefing on the anthrax Investigation: Opening statement by Dr. Vahid Majidi. Available at http://www.fbi.gov/page2/august08/anthraxscience_081808.html.

Federation of American Scientists. 2008a. Unexpected Experimental Results in Transgenic Research: Mousepox Case Study. Available at http://www.fas.org/biosecurity/education/dualuse/FAS_Jackson/index.html.

Federation of American Scientists 2008b. Genomics and Biological Weapons: Poliovirus Case Study. Available at http://www.fas.org/biosecurity/education/dualuse/FAS_Wimmer/FAS_Topic_3_A.html.

Fischer, J. E. 2006. Stewardship or Censorship: Balancing Biosecurity, the Public's Health, and the Benefits of Scientific Openness. Washington, DC: Stimson Center. Available at http://www.stimson.org/globalhealth/pdf/Stewardship.pdf.

Fowler, F. J., Jr. 2001. *Survey Research Methods*, 3rd ed. Thousand Oaks, CA: Sage Publications.

Fox, J. L. 2003. Bioterrorism threat could make some research too "sensitive" to disclose. *ASM News*. March. Available at: http://www.asm.org/microbe/index.asp?bid=13147.

Frerichs, R. L., R. M. Salerno, K. M. Vogel, N. B. Barnett, J. Gaudioso, L. T. Hickok, D. Estes, and D. F. Jung. 2004. Historical Precedence and Technical Requirements of Biological Weapons Use: A Threat Assessment. SAND2004-1854. Albuquerque, NM: Sandia National Laboratories.

Garfinkel M. S., D. Endy, G. L. Epstein, and R. M. Friedman. 2007a. Synthetic genomics: Options for governance. *Industrial Biotechnology* 3(4):333–365.

Garfinkel, M. S., D. Endy, G. L. Epstein, and R. M. Friedman, eds. 2007b. Working Papers for Synthetic Genomics: Risks and Benefits for Science and Society. Available at http://dspace.mit.edu/handle/1721.1/39658.

Gibbons, M. 2006. The year in numbers. In *2005 Profiles of Engineering and Engineering Technology Colleges*. Washington, DC: American Society for Engineering Education.

Greenberg Quinlan Rosner. 2007. Dual Use Research. Washington, DC: GQR Research, Inc.

Gronvall, G. K. 2008. Improving the select agent program. *Bulletin of the Atomic Scientists*. Web ed. October 29. Available at http://www.thebulletin.org/web-edition/features/improving-the-select-agent-program.

Groves, R. M. 2006. Nonresponse rates and nonresponse bias in household surveys. *Public Opinion Quarterly* 70:646–675.

Hansen, S., A. Brewster, J. Asher, and M. Kisielewski. 2006. The Effects of Patenting in the AAAS Scientific Community. Available at: http://sippi.aaas.org/survey/AAAS_IP_Survey_Report.pdf.

Harris, E. D. 2007. Dual use biotechnology research: The case for protective oversight. In *A Web of Prevention: Biological Weapons, Life Sciences and the Governance of Research*, B. Rappert and C. McLeish, eds. London: Earthscan Publications.

Jackson, R. J., A. J. Ramsay, C. D. Christensen, S. Beaton, D. F. Hall, and I. A. Ramshaw. 2001. Expression of mouse interleukin-4 by a recombinant ectromelia virus suppresses cytolytic lymphocyte responses and overcomes genetic resistance to mousepox. *Journal of Virology* 7:1205–1210.

Johnson, C., D. Q. Wilber, and D. Eggen. 2008. Government asserts Ivins acted alone. *Washington Post* (August 7):A01.

Journal Editors and Authors Group (Atlas, R. M., P. Campbell, N. R. Cozarelli, G. Curfman, L. Enquist, G. Fink, A. Flanagin, J. Fletcher, E. George, G. Hammes, D. Heyman, T. Inglesby, S. Kaplan, D. Kennedy, J. Krug, R. Levinson, E. Marcus, H. Metzger, S. S. Morse, A. O'Brien, A. Onderdonk, G. Poste, B. Renault, R. Rich, A. Rosengard, S. Salzburg, M. Scanlan, T. Shenk, H. Tabor, H. Varmus, E. Wimmer, and K. Yamamoto). 2003a. Uncensored exchange of scientific results. *Proceedings of the National Academy of Sciences* 100:1464.

Journal Editors and Authors Group (Atlas, R. M., P. Campbell, N. R. Cozarelli, G. Curfman, L. Enquist, G. Fink, A. Flanagin, J. Fletcher, E. George, G. Hammes, D. Heyman, T. Inglesby, S. Kaplan, D. Kennedy, J. Krug, R. Levinson, E. Marcus, H. Metzger, S. S. Morse, A. O'Brien, A. Onderdonk, G. Poste, B. Renault, R. Rich, A. Rosengard, S. Salzburg, M. Scanlan, T. Shenk, H. Tabor, H. Varmus, E. Wimmer, and K. Yamamoto). 2003b. Statement on the consideration of biodefence and biosecurity. *Nature* 421:771.

Journal Editors and Authors Group (Atlas, R. M., P. Campbell, N. R. Cozarelli, G. Curfman, L. Enquist, G. Fink, A. Flanagin, J. Fletcher, E. George, G. Hammes, D. Heyman, T. Inglesby, S. Kaplan, D. Kennedy, J. Krug, R. Levinson, E. Marcus, H. Metzger, S. S. Morse, A. O'Brien, A. Onderdonk, G. Poste, B. Renault, R. Rich, A. Rosengard, S. Salzburg, M.

Scanlan, T. Shenk, H. Tabor, H. Varmus, E. Wimmer, and K. Yamamoto). 2003c. Statement on scientific publication and security. *Science* 299:1149.

Kahn, L. 2007. Establishing a code of conduct in the life sciences. *Bulletin of the Atomic Scientists.* Web ed. April 1. Available at http://www.thebulletin.org/web-edition/columnists/laura-h-kahn/establishing-a-code-conduct-life-sciences.

Kaiser Family Foundation. 2002. National Survey on the Public's Attitudes Towards HIV/AIDS in the US and the World. June. Available at: http://www.kff.org/kaiserpolls/upload/National-Survey-on-the-Public-s-Attitudes-towards-HIV-AIDS-in-the-U-S-and-the-World-Survey-Toplines.pdf.

Kaplan, S. 2008. The dual-use dilemma: Does it exist? Presentation at ASM Biodefense and Emerging Diseases Research Meeting, February 24, Baltimore, MD.

Keeter, S., C. Miller, A. Kohut, R. M. Groves, and S. Presser. 2000. Consequences of reducing nonresponse in a national telephone survey. *Public Opinion Quarterly* 64:125–148.

Kempner, J., C. S. Perlis, and J. F. Merz. 2005. Ethics: Forbidden knowledge. *Science* 307(5711): 854.

Kennedy, D. 2005. Better never than late. *Science* 310:195.

Khan, M. 2006. Preparations and expectations. Presentation to the United Nations General Assembly First Committee, Sixth Review Conference of the Biological and Toxin Weapons Convention, October 11, New York. Available at: http://www.unog.ch/80256EDD006B8954/(httpAssets)/298DFC7CC2CD636BC125720D0045B3C8/$file/First_Committee_BWC_thematic_presentation_slides.pdf.

Knezo, G. J. 2004. "Sensitive but Unclassified" and Other Federal Security Controls on Scientific and Technical Information: History and Current Controversy. Washington, DC: Congressional Research Service.

Korenman, S. G., R. A. Berk, N. S. Wenger, and V. Lew. 1998. Evaluation of the research norms of scientists and administrators responsible for academic research integrity. *Journal of the American Medical Association* 279:41–47.

Kwik, G., J. Fitzgerald, T. V. Inglesby, and T. O'Toole. 2003. Biosecurity: Responsible stewardship of bioscience in an age of catastrophic terrorism. *Biosecurity and Bioterrorism* 1:27–35.

Leitenberg, M. 1999. Aum Shinrikyo's efforts to produce biological weapons: A case study in the serial propagation of misinformation. *Terrorism and Political Violence* 2(4):149–158.

Leitenberg, M. 2005. Assessing the Biological Weapons and Bioterrorism Threat. Carlisle Barracks, PA: Strategic Studies Institute, U.S. Army War College.

Lentzos, F. 2006. Managing biorisks: Considering codes of conduct. *Nonproliferation Review* 13(2):211–226.

Lichtblau, E. 2001. Response to terror; rising fears that what we do know can hurt us. *Los Angeles Times* (November 18):A1.

Macrina, F. 2007. Scientific societies and promotion of the responsible conduct of research: Codes, policies, and education. *Academic Medicine* 82:865–869.

Martinson, B. C., M. S. Anderson, and R. De Vries. 2005. Scientists behaving badly. *Nature* 435(7043):737–738.

Martinson, B. C., M. S. Anderson, A. L. Crain, and R. De Vries. 2006. Scientists' perceptions of organizational justice and self-reported misbehaviors. *Journal of Empirical Research in Human Research Ethics* 1(1):51–66.

Maurer, S. M., K. V. Lucas, and S. Terrell. 2006. From Understanding to Action: Community-Based Options for Improving Safety and Security in Synthetic Biology. University of California, Berkeley. Available at http://gspp.berkeley.edu/iths/UC%20White%20Paper.pdf.

McCrary, S. V., C. B. Anderson, J. Jakovljevic, T. Khan, L. B. McCullough, N. P. Wray, and B. A. Brody. 2000. A national survey of policies on disclosure of conflicts of interest in biomedical research. *New England Journal of Medicine* 343:1621–1626.

McLeish, C., and P. Nightingale. 2005. Effective Action to Strengthen the BTWC Regime: The Impact of Dual Use Controls on UK Science. Bradford Briefing Paper 17. Available at http://www.brad.ac.uk/acad/sbtwc/briefing/BP_17_2ndseries.pdf.

Merkle, D., and M. Edelman. 2002. Nonresponse in exit polls: A comprehensive analysis. Pp. 243–258 in *Survey Nonresponse*, R. M. Groves, D. A. Dillman, J. L. Eltinge, and R. J. A. Little, eds. New York: Wiley.

Meselson, M. 2000. The problem of biological weapons. Presentation at Symposium on Biological Weapons and Bioterrorism, National Academy of Sciences, Washington, DC, May 2.

MORI. 2000. The Role of Scientists in Public Debate. Prepared for The Wellcome Trust. Available at http://www.wellcome.ac.uk/stellent/groups/corporatesite/@msh_peda/documents/web_document/wtd003425.pdf.

NAS/NAE/IOM (National Academy of Sciences, National Academy of Engineering, Institute of Medicine). 1982. *Scientific Communication and National Security.* Washington, DC: The National Academies Press.

NAS/NAE/IOM. 2002. *Making the Nation Safer: The Role of Science and Technology in Countering Terrorism.* Washington, DC: The National Academies Press.

National Science Board. 2008. *Science and Engineering Indicators 2008.* Volume 1, NSB 08-01; Volume 2, NSB 08-01A. Arlington, VA: National Science Foundation.

Nature. 2005. Special report: The 1918 flu virus is resurrected. 437:794–795.

NRC (National Research Council). 2004a. *Biotechnology Research in an Age of Terrorism.* Washington, DC: The National Academies Press.

NRC. 2004b. *Seeking Security: Pathogens, Open Access, and Genomic Databases.* Washington, DC: The National Academies Press.

NRC. 2006a. *Globalization, Biosecurity, and the Future of the Life Sciences.* Washington, DC: The National Academies Press.

NRC. 2006b. *Policy Implications of International Graduate Students and Postdoctoral Scholars in the United States.* Washington, DC: The National Academies Press.

NRC. 2007a. *Science and Security in a Post 9/11 World: A Report on Regional Discussions Between the Science and Security Communities.* Washington, DC: The National Academies Press.

NRC. 2007b. *Rising Above the Gathering Storm: Energizing and Employing America for a Brighter Future.* Washington, DC: The National Academies Press.

NRC. 2008. *2nd International Forum on Biosecurity: Report of an International Meeting, Budapest, Hungary, March 30–April 2.* Washington, DC. The National Academies Press.

NRC. 2009. *Beyond Fortress America: National Security Controls on Science and Technology in A Globalized World.* Washington, DC. The National Academies Press.

NSABB (National Science Advisory Board for Biosecurity). 2004. Charter: National Science Advisory Board for Biosecurity. Available at http://www.biosecurityboard.gov/SIGNED%20NSABB%20Charter.pdf.

NSABB. 2006. Addressing Biosecurity Concerns Related to the Synthesis of Select Agents. Available at http://www.biosecurityboard.gov/pdf/Final%20NSABB%20Report%20on%20Synthetic%20Genomics.pdf.

NSABB. 2007. *Proposed Framework for the Oversight of Dual Use Life Sciences Research: Strategies for Minimizing the Potential Misuse of Research Information.* Available at http://www.biosecurityboard.gov/news.asp.

NSF (National Science Foundation). 2006. *Characteristics of Doctoral Scientists and Engineers in the United States: 2003.* Division of Science Resources Statistics, NSF 06-320. Arlington, VA: NSF.

Orr, R. 2008. Harnessing the benefits of the biotechnology revolution while managing the potential risks: The role of the United Nations. Presentation at Sustaining Progress in the Life Sciences: Strategies for Managing Dual Use Research of Concern. Bethesda, MD, November 6.

Owens, L. K. 2005. *Introduction to Survey Design.* Chicago: Survey Research Laboratory, University of Illinois at Chicago. Available at http://www.srl.uic.edu/SEMINARS/surveydesign.DOC.

Pearson, G. S. 1993. Prospects for chemical and biological arms control: The web of deterrence. *Washington Quarterly* 16(Spring):145–162.

Pearson, G. S. 2004. A Code of Conduct for the Life Sciences: A Practical Approach. Bradford Briefing Paper 15. Available at: http://www.brad.ac.uk/acad/sbtwc/briefing/BP_15_2ndseries.pdf.

Petro, J. B. 2007. Intelligence Support to the Life Science Community: Mitigating Threats from Bioterrorism. Available at https://www.cia.gov/library/center-for-the-study-of-intelligence/csi-publications/csi-studies/studies/vol48no3/article06.html.

Price, J. H., J. A. Dake, and R. Islam. 2001. Selected ethical issues in research and publication: Perceptions of health education faculty. *Health Education & Behavior* 28(1):51–64.

Rabino, I. 1998. Ethical debates in genetic engineering: U.S. scientists' attitudes on patenting, germ-line research, food labeling, and agri-biotech issues. *Politics and Life Sciences* 17(2):147–163.

Rabino, I. 2003a. Genetic testing and its implications: Human genetics researchers grapple with ethical issues. *Science Technology and Human Values* 28:365–402.

Rabino, I. 2003b. Gene therapy: Ethical issues. *Theoretical Medicine and Bioethics* 24(1):31–58.

Rappert, B. 2003a. Biological weapons, genetics, and social analysis: Emerging responses, emerging issues, II. *New Genetics and Society* 22:297–314.

Rappert, B. 2003b. Coding ethical behavior: The challenges of biological weapons. *Science and Engineering Ethics* 9:453–470.

Rappert, B. 2004. Towards a Life Science Code: Countering the Threats from Biological Weapons. Bradford Briefing Paper 13. Available at http://www.brad.ac.uk/acad/sbtwc.

Rappert, B. 2007. *Biotechnology, Security, and the Search for Limits: An Inquiry into Research and Methods.* Houndmills, UK: Palgrave Macmillan.

Rappert, B. 2008. The benefits, risks, and threats of biotechnology. *Science & Public Policy* 35(1):37–44.

Rappert, B., M. Chevrier, and M. Dando. 2006. In-Depth Implementation of the BTWC: Education and Outreach. Bradford Review Conference Paper 18. Available at http://www.brad.ac.uk/acad/sbtwc/briefing/RCP_18.pdf.

Revill, J., and M. R. Dando. 2006. A Hippocratic oath for life scientists. *EMBO Reports* 7(Special Issue):S55–S60.

Revill, J., and M.R. Dando. 2008. Life scientists and the need for a culture of responsibility: after education . . . what? *Science and Public Policy* 35(1):29–35.

Rice, C. 2001. Letter to Dr. Harold Brown from Condoleezza Rice, Assistant to the President for National Security Affairs. November 1. Available at http://www.fas.org/sgp/bush/cr110101.html.

Roberts, B. (ed.). 2000. *Hype or Reality: The "New Terrorism" and Mass Casualty Attacks.* Alexandria, VA: CBACI.

Royal Society. 2008. Royal Society activities on reducing the risk of the misuse of scientific research. Available at: http://royalsociety.org/document.asp?tip=1&id=7964.

Royal Society/Research Councils UK/The Wellcome Trust. 2006. Factors Affecting Science Communication. Available at http://www.royalsoc.ac.uk/downloaddoc.asp?id=3048.

Salyers, A. 2002. Science, censorship, and public health. *Science* 296(5568):617.

Schemo, D. J. 2002. After 9/11, universities destroying bio agents. *New York Times* (December 17):A20.

Schonlau, M., R. D. Fricker, Jr., and M. N. Elliott. 2002. *Conducting Research Surveys via E-mail and the Web.* Santa Monica, CA: RAND Corporation.

Selgelid, M. J. 2007. A tale of two studies. Ethics, bioterrorism, and the censorship of science. *Hastings Center Report* 37(3):35–43.

Shane, S., and E. Lichtblau. 2008. Scientist's suicide linked to anthrax inquiry. *New York Times.* (August 2). Available at http://www.nytimes.com/2008/08/02/washington/02anthrax.html#.

Sharp, P. A. 2005. 1918 Flu and responsible science. *Science* 310(5745):17.

Shea, D. 2007. Oversight of Dual-Use Biological Research: The National Science Advisory Board for Biosecurity. Washington, DC: Congressional Research Service.

Simon, J., and M. Hersh. 2002. An educational imperative: The role of ethical codes and normative prohibitions in CBW-applicable research. *Minerva* 40:37–55.

Smith, H. O., C. A. Hutchison III, C. Pfannkoch, and J. C. Venter. 2003. Generating a synthetic genome by whole genome assembly: Phi X174 bacteriophage from synthetic oligonucleotides. *Proceedings of the National Academy of Sciences* 100: 15440–15445.

Somerville, M. A., and R. M. Atlas. 2005. Ethics: A weapon to counter bioterrorism. *Science* 307(5717):1881–1882.

Stebbins, M. 2008. The "What if?" of Dual-Use Research Awareness: How Should Scientists Deal with Their Concerns? Scientists and Engineers Action Fund Blog. Available at http://sefora.org/2008/06/11/the-%e2%80%9cwhat-if%e2%80%9d-of-dual-use-research-awareness/.

Steinbruner, J. D., E. D. Harris, N. Gallagher, and S. M. Okutani. 2007. *Controlling Dangerous Pathogens.* College Park, MD: Center for International Security Studies at Maryland.

Sutton, V. 2007. Biodefense Researchers Survey of Opinions on the Select Agent Regulations and Achieving the Goal of National Security and Protecting Public Health. Center for Biodefense, Law and Public Policy, Texas Tech University School of Law.

Sutton, V. 2009. Letter to the Editor: Survey Finds Biodefense Researcher Anxiety Over Inadvertently Violating Regulations. *Biosecurity and Bioterrorism: Strategy, Practice, and Science* 7(2):forthcoming.

Taubenberger, J. K., A. H. Reid, R. M. Lourens, R. Wang, G. Jin, and T. G. Fanning. 2005. Characterization of the 1918 influenza virus polymerase genes. *Nature* 437:889–893.

Timmer, J. 2008. AAAS: Ethics in scientific publishing. Available at http://arstechnica.com/journals/science.ars/2008/02/17/aaas-ethics-in-scientific-publishing.

Tumpey, T. M., C. F. Basler, P. V. Aguilar, H. Zeng, A. Solórzano, D. E. Swayne, N. J. Cox, J. M. Katz, J. K. Taubenberger, P. Palese, and A. García-Sastre. 2005. Characterization of the reconstructed 1918 Spanish influenza pandemic virus. *Science* 310:77–80.

Tucker, J. (ed.). 2000. *Toxic Terror: Assessing Terrorist Use of Chemical and Biological Weapons.* Cambridge, MA: MIT Press.

Union of Concerned Scientists. 2005a. NOAA Fisheries Survey Summary. Available at http://www.ucsusa.org/scientific_integrity/interference/survey-political-interference-at-noaa-fisheries.html.

Union of Concerned Scientists. 2005b. U.S. Fish and Wildlife Service Survey Summary. Available at http://www.ucsusa.org/scientific_integrity/interference/us-fish-wildlife-service-survey.html.

Union of Concerned Scientists. 2006. Food & Drug Administration Scientist Survey. Available at http://www.ucsusa.org/scientific_integrity/interference/fda-scientists-survey-summary.html.

Union of Concerned Scientists. 2008. Environmental Protection Agency Survey. Available at http://www.ucsusa.org/scientific_integrity/interference/interference-at-the-epa.html.

United Nations. 2006. The United Nations Global Counter-Terrorism Strategy. UNGA Resolution A/RES/60/288. New York: United Nations, II/11. Available at http://www.un.org/terrorism/strategy-counter-terrorism.shtml#resolution.

U.S. Department of State. 2006. Renewal of UN Security Council Resolution 1540 Committee Mandate. June 9. Available at http://www.state.gov/t/isn/rls/other/67974.htm.

U.S. Government. 2008. Outreach and Education in the Life Sciences: Case Study in the U.S. Department of Energy National Laboratories. Submitted to the 2008 Meeting of Experts of the Biological Weapons Convention, Geneva, Switzerland. Available at http://daccessdds.un.org/doc/UNDOC/GEN/G08/621/06/PDF/G0862106.pdf?OpenElement.

Van Aken, J. 2006. When risk outweighs benefit. *EMBO Reports* 7(Special Issue):S10–S13.

Vastag, B. 2002. Openness in biomedical research collides with heightened security concerns. *Journal of the American Medical Association* 289:686–690.

Venter, J. C. 2007. *A Life Decoded: My Genome, My Life.* New York: Viking.

Vogel, K. M. 2008. Framing biosecurity: An alternative to the biotech revolution model? *Science and Public Policy* 35(1):45–54.

Vogeli, C., R. Yucel, E. Bendavid, L. M. Jones, M. S. Anderson, K. S. Louis, and E. G. Campbell. 2006. Data withholding and the next generation of scientists: Results of a national survey. *Academic Medicine* 81(2):128–136.

von Bubnoff, A. 2005. The 1918 flu virus is resurrected. *Nature* 437:794–795. Available at http://www.nature.com/nature/journal/v437/n7060/full/437794a.html.

Wein, L. M., and Y. Liu. 2005. Analyzing a bioterror attack on the food supply: The case of botulinum toxin in milk. *Proceedings of the National Academy of Sciences* 102(28):9984–9989.

Weisberg, H.F. 2005. *The Total Survey Error Approach: A Guide to the New Science of Survey Research.* Chicago: University of Chicago Press.

Wenger, N. S., S. G. Korenman, R. Berk, and S. Berry. 1997. The ethics of scientific research: An analysis of focus groups of scientists and institutional representatives. *Journal of Investigative Medicine* 45:371–380.

Wenger, N. S., S. G. Korenman, R. Berk, and H. Liu. 1999. Reporting unethical research behavior. *Evaluation Review* 23:553–570.

White House. 1969. United States Policy on Chemical Warfare Program and Bacteriological/Biological Research Program (NSDM 35). Available at http://www.state.gov/documents/organization/90919.pdf.

White House. 1985. National Policy on the Transfer of Scientific, Technical and Engineering Information (NSDD-189). Available at http://www.fas.org/irp/offdocs/nsdd/nsdd-189.htm.

White House. 2008. Designation and Sharing of Controlled Unclassified Information (CUI). Office of the Press Secretary (May 7). Available at http://www.fas.org/sgp/bush/cui.html.

White House 2009. Strengthening Laboratory Biosecurity in the United States. Executive Order (January 9). Available at http://74.125.47.132/search?q=cache:ZXz3J17R4vEJ:https://www.whitehouse.gov/news/releases/2009/01/20090109-6.html+%22Strengthening+Laboratory+Biosecurity+in+the+United+States+%22&hl=en&gl=us&strip=1.

WHO (World Health Organization). 2005. Life Science Research: Opportunities and Risks for Public Health. Geneva: WHO. Available at http://www.who.int/ethics/Life%20Science%20Research.pdf.

WHO. 2007. Scientific Working Group on Life Science Research and Global Health Security: Report of the First Meeting. Geneva: WHO. Available at http://www.who.int/csr/resources/publications/deliberate/WHO_CDS_EPR_2007_4.

Willman, D., and D. G. Savage. 2008. FBI says evidence points uniquely to Bruce Ivins in anthrax case. *LA Times* (August 7). Available at http://www.latimes.com/news/printedition/front/la-na-anthrax7-2008aug07,0,491144.story.

Wimmer, E. 2006. The test-tube synthesis of a chemical called poliovirus. *EMBO Reports* 7(Special Issue):S3–S9.

Wright, S. 2007. Terrorists and biological weapons: Forging the linkage in the Clinton administration. *Politics and the Life Sciences* 25(1–2):57–115.

Zilinskas, R., and J. B. Tucker 2002. Limiting the contribution of the open scientific literature to the biological weapons threat. *Online Journal of Homeland Security* (December). Available at http://www.homelandsecurity.org/journal/Articles/.tucker.html.

Appendixes

Appendix A

Committee Member Biographies

Ronald M. Atlas (*Chair*), professor of biology and public health, and co-director of the Center for the Health Hazards Preparedness at the University of Louisville. He received his B.S. degree from the State University of New York at Stony Brook and his M.S. and Ph.D. degrees from Rutgers University. He was a postdoctoral fellow at the Jet Propulsion Laboratory where he worked on Mars life detection. He is chair of the National Aeronautics and Space Administration's Planetary Protection Board, chair of the Wellcome Trust Infection, Immunology, and Population Health Strategy Committee, and and co-chair of the American Society for Microbiology (ASM) Task Force on Biodefense. He has previously served as president of the ASM. He has also previously served on the National Institutes of Health (NIH) Recombinant DNA Advisory Committee, the Department of Homeland Security Science and Technology Advisory Committee, and the Federal Bureau of Investigation Scientific Working Group on Bioforensics. His early research focused on oil spills, and he discovered bioremediation as part of his doctoral studies. Later he turned to the molecular detection of pathogens in the environment, which forms the basis for biosensors to detect biothreat agents. He is author of nearly 300 manuscripts and 20 books. He is a fellow in the American Academy of Microbiology and has received the ASM Award for Applied and Environmental Microbiology, the ASM Founders Award, the Edmund Youde Lectureship Award in Hong Kong, and an honorary doctor of science degree from the University of Guelph.

Robert Cook-Deegan has been the Director of the Institute for Genome Sciences and Policy Center for Genome Ethics, Law, & Policy at Duke University since July 2002. Prior to coming to Duke, he was director of the Robert Wood Johnson Foundation Health Policy Fellowship program at the Institute of Medicine (IOM), National Academy of Sciences (NAS). Dr. Cook-Deegan was a Cecil and Ida Green Fellow at the University of Texas, Dallas, following his work on the report *Allocating Federal Funds for Science and Technology* (the "Press report"). From 1991 through 1994, he directed IOM's Division of Biobehavioral Sciences and Mental Disorders (since renamed Neuroscience and Behavioral Health). He worked for the National Center for Human Genome Research (1989–1990), after serving as acting executive director of the Biomedical Ethics Advisory Committee of the U.S. Congress (1988–1989). He is the author of *The Gene Wars: Science, Politics, and the Human Genome*. Dr. Cook-Deegan was a congressional science fellow in 1982 and spent 5 years at the congressional Office of Technology Assessment. Dr. Cook-Deegan did 2 years of postdoctoral research on the molecular biology of oncogenes with Lasker Award scientist Raymond L. Erikson, after completing his internship in pathology at the University of Colorado (1979–1982). He received his bachelor's degree in chemistry, magna cum laude, in 1975 from Harvard College, and his M.D. degree from the University of Colorado in 1979. He is secretary and trustee of the Foundation for Genetic Medicine. Dr. Cook-Deegan was a member of the Board of Directors, Physicians for Human Rights (1988–1996), with whom he participated in human rights missions to Turkey, Iraq, and Panama.

David Franz is vice president and chief biological scientist, Midwest Research Institute; director, National Agricultural Biosecurity Center, Kansas State University; and deputy director, Center for Emergency Care and Preparedness, University of Alabama at Birmingham. Dr. Franz served in the U.S. Army Medical Research and Materiel Command for 23 of 27 years on active duty and retired as colonel. He served as commander of the U.S. Army Medical Research Institute of Infectious Diseases and as deputy commander of the Medical Research and Materiel Command. Prior to joining the command, he served as group veterinarian for the 10th Special Forces Group (Airborne). Dr. Franz was the chief inspector on three UN Special Commission biological warfare inspection missions to Iraq, and served as technical adviser on long-term monitoring. He also served as a member of the first two U.S.–UK teams that visited Russia in support of the Trilateral Joint Statement on Biological Weapons and as a member of the Trilateral Experts' Committee for biological weapons negotiations. Dr. Franz was technical editor for the *Textbook of Military Medicine on Chemical and Biological Defense* released in 1997. Current committee appointments

include the Defense Intelligence Agency Red Team Bio-Chem 2020, the Defense Threat Reduction Agency's Threat Reduction Advisory Committee, the NAS Committee for Research with Russian Biological Institutes (chair), the NAS Committee on International Security and Arms Control, the Department of Health and Human Services' National Science Advisory Board for Biosecurity, and the recently decommissioned Department of Homeland Security Science and Technology Advisory Committee. Dr. Franz holds an adjunct appointment as professor for the Department of Diagnostic Medicine and Pathobiology at the College of Veterinary Medicine, Kansas State University, and serves on the Dean's Advisory Council. He also holds an adjunct appointment as professor in the Department of Emergency Medicine at University of Alabama at Birmingham. Dr. Franz serves as a senior fellow in the Combating Terrorism Center of the U.S. Military Academy at West Point. Dr. Franz holds a D.V.M. from Kansas State University and a Ph.D. in physiology from Baylor College of Medicine.

James Lepkowski is a research professor at the Institute for Social Research, where he conducts survey methodology research and directs the Summer Institute in Survey Research Techniques. He is a member of the faculty of the Joint University of Maryland–University of Michigan Program in Survey Methods, and is a professor in the Department of Biostatistics at the University of Michigan. Dr. Lepkowski received his Ph.D. from the University of Michigan in 1980. Since that time, he has worked at the Institute for Social Research designing, conducting, analyzing, and evaluating a variety of survey samples, including area probability and telephone samples of households. The substantive content of most of this work has been health or conditions that occur infrequently in the population. Dr. Lepkowski also has conducted investigations into a wide variety of survey methodology problems, including the design of telephone samples for households in the United States, the behavior of analytical statistics when the data are obtained from complex sample surveys, imputation methods to compensate for item missing data in surveys, weighting to compensate for unit nonresponse, and the interaction between interviewer and respondent in the survey interview. He has served on a variety of national and international advisory committees on survey research methods, including service to the World Health Organization, the National Center for Health Statistics, the Food and Drug Administration, the Bureau of the Census, the Bureau of Labor Statistics, and other federal statistical agencies. He is an active member of the American Statistical Association, serving in various offices in the Survey Research Methods Section and on association committees, is a fellow of the association, and an elected member of the International Statistical Institute.

Francis Macrina is vice president for research and Edward Myers Professor at the Philips Institute of Oral and Craniofacial Molecular Biology at Virginia Commonwealth University. His honors include an NIH Research Career Development Award, the Virginia Outstanding Scientist Award, and the NIH NIDCR MERIT Award. His research focuses on human oral microbes. In addition, he has conducted educational research on the effectiveness of formal training in research ethics and is well known for the text *Scientific Integrity: Text and Cases in Responsible Conduct of Research*, currently in its third edition. He received his B.S. from Cornell University and a Ph.D. in microbiology from Syracuse University.

Kathleen Vogel is an assistant professor at Cornell University, with a joint appointment in the Department of Science and Technology Studies and the Peace Studies Program. Her research interests are biological warfare and bioterrorism; nonproliferation and arms control; and military technology and technology transfer. Before coming to Cornell, Dr. Vogel worked with the U.S. Department of State as a William C. Foster Fellow in the Bureau of Nonproliferation in the Office of Proliferation Threat Reduction. She holds a Ph.D. in chemistry from Princeton University. Her current research explores the technical and social factors influencing the proliferation of biological weapons technology to terrorist groups and countries of proliferation concern. At Cornell, Dr. Vogel teaches "The Military and New Technology," which analyzes technological innovation in the military; "The Dark Side of Biology: Biological Weapons, Bioterrorism, and Biocriminality," which examines various analytical frameworks for evaluating biological weapons threats; and "Ethical Issues in Health and Medicine," which explores ethical dilemmas and frameworks in the practice of medicine and the life sciences. Relevant recent publications include "Conversion at Stepnogorsk: What the Future Holds for Former Bioweapons Facilities" (Peace Studies Program Occasional Paper 2003); "Bioweapons Proliferation: Where Science Studies and Public Policy Collide" (*Social Studies of Science* 36:659–690, 2006).

Appendix B

Focus Group Results

GREENBERG QUINLAN ROSNER RESEARCH

March 15, 2007

Dual–Use Research

To: John Sislin

From: Anna Greenberg, Jennifer Berktold, and Jessica Keating

Executive Summary

Biological research contributes greatly to society; it helps scientists understand the causes of disease, develop cures, improve agricultural methods and our food supply and the like. However, the same research, applied in a different context, has the capability to damage rather than heal. This notion that the same biological research can be used for good or ill is also known as "dual use" research. Post-9/11 and, in particular, after the anthrax attacks, stakeholders inside and outside of government have called for increased awareness among the life science community and new processes to prevent misapplication of dual use research.

This current research is the first step of a larger project that seeks to better understand life scientists' perceptions of the debates surrounding dual use research and to explore responsibility for preventing dual use research from being misused. We conducted qualitative research to understand the current level of concern surrounding dual use research, to explore life scientists' views about who, if anyone, should shoulder the responsibility of preventing such research from being misused, and to explore the consequences and concerns surrounding greater controls on dual use research.[1]

We found that the scientists we interviewed were themselves familiar with dual use research – some deeply aware – but for many others, knowledge was superficial. Many life scientists are only exposed to dual use issues as part of their job (e.g., if they work with select agents) or by serving on an institutional review committee (IRC), rather than receiving systematic training in graduate school or by employers. If life scientists are not in an academic environment, do not receive federal funding or do not work with select agents, they are unlikely to have significant exposure to the issues surrounding dual use research.

Moreover, we found that there is no real consensus about how much concern the scientific community places on the possible misapplication of dual use research. We found that many scientists, while aware of dual use issues and concerned about the possibility of misuse, do not

[1] For methodological details, please see Appendix A.

www.greenbergresearch.com

Washington, DC
10 G Street NE, Suite 400, Washington, DC 20002
Phone: 202 478 8300 / Fax: 202 478 8301

California
50 California Street, Suite 1500, San Francisco, CA 941
Phone: 415 277 5403 / Fax: 415 358 9599

worry about the potential for misuse in their own work. Many need to be convinced that regulation of dual use research is necessary at all. And many are content to continue to rely on a patchwork of self-regulation by scientists in labs, oversight by academic IRCs and reviews by federal grantors.

That lack of consensus is rooted in views about the role science plays in society and the scientists' tolerance for risk. The groups indicated that people who were less worried about misuse held deeply the values of transparency, open communication, and the belief that science benefits from unintended consequences. Some also hold an "anything goes" view of research and are unwilling to cede any paths of research from going unstudied. Life scientists concerned about dual use are more likely to see many vulnerabilities that could be easily exploited, with no real mechanisms in place to address them. Because of the potential danger it presents, they hunger for a solution which they hope can save lives and prevent more fear.

But even for life scientists who feel that misuse is a pressing concern, many are reluctant to cede core scientific values like transparency, open flow of information, and a desire to cure diseases, in exchange for added security. The challenge is steep: how to raise awareness and institute practices to reduce the risk of biological research being hijacked for nefarious purposes, while at the same time not hinder the open flow of communication that is necessary to move the field forward.

This memo is based on three focus groups with life scientists from a wide variety of backgrounds. We spoke with participants from academia, government, and private industry, and with expertise in many fields including agriculture, defense, public health, and medical research.

Key Findings

- Unless scientists deal with dual-use research as part of daily life, dual-use issues are not top of mind. Dual use research is understood narrowly by scientists, often in the context of their own jobs. Using a narrow lens, many scientists do not recognize that their own work could be considered dual use research.

- Defining dual use research is challenging. There is a tension between elevating the scope of risk and introducing a definition that is so broad that it diminishes the urgency because almost anything could be defined as dual use. In fact, in the focus groups, a more expansive definition (i.e., NSABB's dual use research of concern) received pushback from respondents because it was seen as too broad and too widely applied.

- Many scientists believe that scientists bear the ultimate burden of responsibility to prevent misuse of dual-use research. If greater controls are needed, they would prefer to see them coming from the science community rather than government. Scientists worry greatly that added regulation will create an operating environment unfriendly to research, in part because rules could be capricious, random or burdensome.

- More scientists are comfortable with procedures that are already in place such as monitoring by principal investigators in labs, oversight by Institutional Review Committees and internal editorial review in the publication process, though many acknowledge that these procedures are haphazard or inconsistent from one to another.

- Developing policies to address dual use issues also presents challenges, because scientists worry greatly about whether such policies will have the unintended consequence of restricting scientific freedom. Underlying this discomfort is a lively debate among scientists on whether there should ever be limits to research. One school of thought thinks of the dangers of reintroducing a harmful agent, like smallpox, into the environment or doing research with the intent to kill; this group is more comfortable in adopting policies to prevent misuse. Another school of thought reflects upon core science values – such as free speech, open mindedness, and the law of unintended consequences - arguing that one can never know where the next scientific breakthrough will come from; this group has trepidation over policies to prevent dual-use research.

Recognition of Dual-Use Research

In the focus groups, all participants were familiar with dual-use research and when asked, the groups could define it correctly. In fact, many of our scientists have a great deal of exposure to dual use research issues because of the work they do; many engage in true dual use research and some were leaders in their fields. But some scientists' understanding of dual-use is narrow – "things related to threat agent pathogens, bioterrorist pathogens." When we introduced a broad definition of dual use research (i.e. the NSABB definition of "dual use research of concern"), some realized that perhaps their own work could be considered dual use. At the same time, this broad definition was easily dismissed because then everything, including "Velcro" or "Viagra," has the potential for misuse.

> ***I mean, I do [dual use research] by this definition and I wouldn't think, if I find out how people can be susceptible to things, that would fit this definition. If I find out differences about gene frequencies in different ethnic groups, that would fit this definition. But if you asked me prior to reading this, I'd say no, I don't do dual-use type work.***
> A Life Scientist in Bethesda, Maryland

> ***I would say [the NSABB definition is] too broad because it's probably language that will go into some regulation and so that's where this stuff falls down.***
> A Life Scientist in Bethesda, Maryland

> ***The problem with that is that it defines virtually all technology, like Velcro.***
> A Life Scientist in Bethesda, Maryland

Though all scientists in our groups were, themselves, familiar with the term "dual-use research," many feel that others in the community do not become familiar with the issues regarding potential misuse unless they encounter it as part of their job. In other words, many recognize there is no systematic training in graduate school or by employers, and life scientists only encounter the issue when they apply for federal grants or when they work in an academic environment.

No one ever learns [about dual use] unless they directly have any obligation to it. Again, I would say it's less than one percent.
A Life Scientist in San Francisco, California

Credibility and Urgency of the Threat

Most scientists agree that biological research, if misused, presents credible and significant threat. Life scientists can easily imagine circumstances under which research can be misused and cause substantial harm. They can also easily recall examples where humans introduced biological agents into the environment, with great impact.

Human viruses... it could be easily aerosolized and just come one day to somewhere like New York or DC, like ebola type or something like that.
A Life Scientist in Bethesda, Maryland

I would worry most about a vaccine resistant smallpox. Actually has the potential to extinguish our nation as it exists, and we really don't have a way to counter it.
A Life Scientist in Bethesda, Maryland

Look at the mileage that whoever portrayed, you know, pulled off the anthrax events got, the publicity involved and the fear that was instilled in the country. It was very small numbers of casualities considering, but still...

17 affected.

17 that were seen by the CDC.

But still, it's just, you know, the fear factor is enormous there.
A Life Scientist in Bethesda, Maryland

However, there is little consensus, even among scientists who engage in dual-use research, about the urgency of the threat. Except for the few scientists actively engaged in research with select agents or with panels that are attempting to find solutions on this issue, the potential for misuse simply is not top of mind for many life scientists.

When I was in academia, I remember going down to Caltech and talking with the professors there. And they all were very much aware of those who I was speaking with of some of these issues, but nobody ever thought that their work could be misused. It just seemed to be, "yes, we know about these things, we talk about the [adrilican four], we talk about [Wimmer's] synthetic polio. But my research couldn't [be misused]." And I remember talking about the grad students who, a couple were friends of mine at that time, and it just didn't really seem to be an issue. Now that was probably three or four years ago.
A Life Scientist in Bethesda, Maryland

> ***Well, I think rather, my own frank opinion of this is rather too much of this has been made out of this issue. Virtually all of the scientists I work with are extremely honorable individuals with high ethical and moral standards and scarcely need the classes that they're dragged to which try to ram further ethics down their throat.***
> A Life Scientist in San Francisco, California

> ***And I actually don't [worry about it everyday]. I mean, in part because it's not part of my job, so I would worry a lot more if I were you guys who do this for a living. But I just think of it in straight probabilistic terms.***
> A Life Scientist in Bethesda, Maryland

> ***In case of large, or not even a large, even a small outdoor release that maybe killed a couple of people, let's say, 10, 20, is going to have a large impact, because we don't have a lot of the policy questions answered. We don't know when it's safe to reoccupy a building. We don't know how clean is clean. We don't know whether the strategy is to treat people in the most medically expeditious way or reduce the risk from contamination. Like if this hospital is potentially in a zone, do you use it? Do you not use it? We haven't addressed any of these policy questions. And because of that, any kind of release that causes unknown contamination over an area is going to be a huge problem. If you evacuate all of downtown Manhattan for two weeks?***
> A Life Scientist in Bethesda, Maryland

Part of the skepticism stems from the fact that other forms of terrorism are easier to execute. These scientists note that bioterrorism requires access to select agents, scientists with the skill to formulate the agent and a means for dissemination. Damage from biological agents often occurs at a slower pace, affording an opportunity to counteract the effects through antibiotics or vaccines. And even if the method of dissemination is simple (such as using an aerosol, scattering it at a restaurant, or releasing it into Times Square), to date the combination of intent and motive has not been present. Indeed, some participants feel that the misapplied biological agent is more likely to accidentally escape, rather than be intentionally placed into the environment.

> ***They've been so successful with explosives. Explosives are reliable.***
> A Life Scientist in Bethesda, Maryland

> ***I think that, [...] for whatever reason, there hasn't yet been the nexus of motive and opportunity and, you know, just, but there have been calls to, for scientists to join, biological scientists to join Al-Qaeda, but for whatever reason, it hasn't yet had that nexus.***
> A Life Scientist in Bethesda, Maryland

> ***I mean, there's three aspects of the problem. There's the agent, formulating the agent, and then there's putting into ignition, weaponization. And I doubt you'd find one person that's good at all three.***
> A Life Scientist in Bethesda, Maryland

> ***I thought that there were just easier ways for a group to engage in acts of terrorism than trying to find disgruntled scientists or taking the time to grow up things. I mean, for a number of the biological agents, you also have treatments that are effective. There's some vaccines that are available. You know, biologicals aren't quite the same as [nukes or radiation where] you only have minutes to distribute medical countermeasures, whereas biologics you might have a little bit more time. I just, I don't know, I'm not as concerned with biological agents as I am with some other things [...].***
> A Life Scientist in Bethesda, Maryland

> ***I think there's a great deal of false hubris associated with anybody who thinks they can deal with a pathogen that nature has had millions of years of development and experience. So I find hollow comments that somebody can build a more serious pathogen than nature has had the opportunity to. And I agree with [others'] comments that, one, I think that terrorism is more likely to be nonbiological over the next five years, and then, secondly, I think if it were to be biological, it's likely to use one of the gold standard pathogens.***
> A Life Scientist in San Francisco, California

Of course, there are important variations among life scientists. For instance, our research suggests there are substantial regional differences in the level of concern regarding misuse of dual use research. Our participants outside of Washington D.C. simply did not think that misuse is a major concern; instead they prioritize finding solutions regardless of whether the trigger is natural, accidental or intentional. Scientists inside the beltway are much more sensitized to this issue, perhaps because the city has been directly affected by terrorism, and perhaps due to the composition of the focus groups, which included people who work in government agencies.

> ***Yeah, it would be interesting to see if scientists in Washington are far more aware versus scientists in Idaho. I bet you'll find them a lot less aware of the dual use term.***
> A Life Scientist in Bethesda, Maryland

> ***Doesn't everybody have that neighbor at this point, where their consciousness has been raised by the very visible cases and the fact that they know there is a government oversight committee to try and develop some guidelines that everybody would use.***
> A Life Scientist in Bethesda, Maryland

> ***I think it's absolutely true that there are dangers posed by pathogens. Whether those pathogens are produced at the hand of man or whether those pathogens are produced by nature, and, by the way, the latter is much more likely in the foreseeable future, they still could pose a threat to us, and we'd defend against them in largely the same ways.***
> A Life Scientist in San Francisco, California

Finally, for some more liberal scientists, it is difficult to separate their overall criticism of the Bush Administration's "war on science" and prosecution of the war on terrorism from the dual use issue. In particular, for participants outside the beltway, many wondered whether misuse of

dual use research presents a legitimate threat or if it is simply an exaggeration from this administration.

> ***We've been taught to have this threat, we've been brainwashed to have this threat by this current administration, and that if, again, if we had a completely different persona in the world, we would feel a lot more secure. So rather than just thinking in terms of defense for bioterrorism, this whole idea of turning around the image and the sort of, it sounds like a little bit of a polemic here, but I think that we're much less secure as a result of our foreign policy and the way we put ourselves in the world as the United States.***
> A Life Scientist in San Francisco, California
>
> ***Oh, I'm so tired of people going before congressmen, or the congressmen getting up and saying, this person is providing a blueprint for terrorism for publishing this paper on reverse genetics and HIV or reverse genetics and flu or something. It's such crap. Probably everybody at the table feels this way.***
> A Life Scientist in San Francisco, California
>
> ***It's hard to get the government to respond to the advice of scientists.***
>
> ***It's the other way around. The government doesn't get the information from the scientists. The scientists know what's going on in the halls. [...]***
>
> ***There's a war on science.***
> Life Scientists in San Francisco, California

Science Values and Dual Use Research

For many scientists, the values that they hold dear to promote good science are antithetical to regulation that prevents misuse. Life scientists value an open marketplace of research and exchange of ideas that lead to advances in the field. This dialogue requires transparency of research processes and publication of research relevant to their own field of specialty. Moreover, as biologists, they are drawn to the field by a desire to help people, believe they are promoting the common good, and are inclined to believe that their compatriots are in it for the right reasons. Trading these values off against the potential harm that can come from misapplied research, there is a lively debate among these scientists about whether it is ever appropriate to limit the scope of life science research.

Open flow of information is crucial to scientific research. Of course, the "publish or perish" imperative is a professional requirement for many, but publishing is also viewed as necessary to further the field. Scientists are uneasy with the notion that restrictions could be put in place in that would, in effect, limit what research can be shared. Transparency is necessary to disseminate findings and being able to replicate research. Even scientists very concerned with dual use research are hesitant to employ measures that would limit publication or other means of communicating discoveries.

> ***If you don't communicate the research, it's not complete... if the whole point is to advance, whether it's knowledge or, in my case, to promote an idea or to suggest improvement to a system, it's not, it doesn't, you know, you can talk to yourself all day long, but it's more helpful to talk to other people.***
> A Life Scientist in Bethesda, Maryland

> ***And the more open we are about things. I frankly believe, I believe in security through transparency rather than security through obscurity. And I think there's a lot of support for that in the community that really knows about this. Not so much support in the policy making communities, not in the halls of Washington. But if you talk amongst the people who really understand this problem who've worked it their whole lives, you'll discover that they have very different attitudes about how you could become more secure.***
> A Life Scientist in San Francisco, California

It is unthinkable for many of these scientists to sacrifice flow of information and scientific advancement in exchange for security for "protection" from misuse particularly since there is no agreement on what it means to protect research from misuse. Often this happens because in the context of the focus group they are asked to describe behavior that originates from common sense or instinct, rather than a prescribed set of procedures. Moreover, their hesitancy towards added protections are tied to two separate values of science. First, scientists feel they need to be able to take risks in order to make discoveries. Second, it is always unclear where discoveries come from – often they are a result of unintended consequences.

> ***The arrogance of people that think they can figure out what's going to happen with experiments blows my mind. Science that does experiments to try to get a result if you think you know the answer are worthless. It's the ability to do experiments where you don't know what the answer is, you look for the opposite, you try to figure out what that means, and that requires this kind of risk taking. You can't anticipate what you're going to find.***

> ***That's a good point. Scientists are risk takers, and unless we're prepared to allow a level of risk, there's going to be no benefit that comes out of all this stuff. You hit the nail on the head. And all this legislation, well intentioned as it may be, goes back to this law of unintended consequences. The unintended consequence is to shut down the research part of research. And what we're going to be left with is nothing that's going to be useful, particularly if you do believe that there's looming in the future some danger posed to mankind by either the next really bad emerging disease or, God forbid, a genetically engineered pathogen.***
> Two Life Scientists in San Francisco, California

Beyond concerns like transparency and the need to publish, biology – as a discipline - has qualities that make it exceptional. People enter into careers in biology to make others healthy, to cure diseases. Biologists assume this about one another, which makes it difficult to perceive circumstances under which a biologist would want to use their knowledge for terrorist purposes - to harm rather than heal.

Biology is somewhat different from other fields, because the people in it have always felt that they're in it for beneficial, healthy, altruistic reasons, and they're feeling that, you know, they didn't start out to make a bomb. They started out to find ways to defeat illness [...]. So they would be perhaps more likely, by matter of choice, to not see the, to not want to see the malevolent attempt to put their research to.
A Life Scientist in Bethesda, Maryland

You've got a lot of better things to do. If you're an expert in munitions or explosives, that's your job. Biologists, there's a lot of biologists, but most of them have no interest or no desire to do it.
A Life Scientist in Bethesda, Maryland

These underlying values underscore what is a real debate, with no consensus, about whether or not there should be limits to life science research. Measures to prevent misuse of dual-use research could have the unintended consequence of closing the door on certain paths of research. While acceptable to some, this possibility is deeply unsettling to others. Some feel that there is never research that is inappropriate for study, while others readily talked about the dangers of reintroducing small pox or scarlet fever to people.

There's a potential that other people might be studying it and it's going to be misused anyway. [...] It's foolish to think that just because researchers in the U.S. aren't doing it that it's not being done someplace else.
A Life Scientist in Bethesda, Maryland

Experiments to weaponize microorganisms with the sole purpose of using them as bio-weapons, I don't think should be done.

So putting multi drug resistance in anthrax would be a no-no.
Two Life Scientists in San Francisco, California

The Onus of Responsibility

Scientists, when forced to focus on responsibility for preventing misuse of dual-use research, believe scientists, not the government, have the primary responsibility to exercise due diligence and prevent misuse. They prefer to rely on self-regulation and believe that their colleagues are trustworthy, though few report that they are pro-active or taking forward-looking steps to prevent misuse. Many are also comfortable with existing procedures, though many would acknowledge there are no centralized procedures in labs, from the government, in journals, and even in institutional review committees (IRCs) regarding dual use research. There is no doubt that, at least in these focus groups, scientist prefer self-regulation to government regulation.

> ***I've always talked about just simple trust. So when I talk to the safety folks in my area of work, I say at the end of the day, you just have to trust that I'm not going to do anything bad. And the people that you allow into these areas, you have to trust us. We've been trained as scientists, we know what's right and what's wrong, and we have to trust that we're going to do what's right.***
> A Life Scientist in San Francisco, California

As a result, any actions currently taken by scientists to prevent misuse are haphazard, not based in established procedure (except when dealt with through academic IRBs or as required by government grants). Not surprisingly, this process is imperfect and incomplete. First, every lab has different rules set by the manager, generally the principal investigator. Thus, there are no consistent policies regarding dual use research from one lab to another or from one government agency to another.

> ***Eventually somebody has to, the head of the lab has to have some responsibility to make sure that he or she knows what's going on inside.***
> A Life Scientist in Bethesda, Maryland

> ***We all have different ideas of what constitutes sterile technique. But if you grew up in a lab that grew E. coli, you may have one idea, which is pretty cavalier. If you grew up with yeast, you also have a pretty cavalier idea. But if you grew up in a lab that did, say, cell cultures right next to the lab that was working with yeast, you'd have a very different idea of what constitutes sterile technique. And if you're working with something that's a human pathogen, you have, yet, again, a heightened awareness of just what it takes to keep these organisms out of the environment in an unwanted way. And so, yes, there are best practices, and those best practices are honed by experience from working with the very agents that those practices were developed for. There's no one set of best practices for any organism, there are multiple sets.***
> A Life Scientist in San Francisco, California

> ***There are NIH guidelines, in my opinion, from my own experience, one of the problems is there are NIH guidelines to do work with pathogens, and then you have CDC in the middle of this which sometimes they don't appear to talk to each other, because they have different requirements, and so forth. So there are guidelines, but I guess they are not standardized.***
> A Life Scientist in San Francisco, California

Participants are very hesitant to see government agencies take a larger role in dual-use issues. Many of these scientists were directly engaged with government research – either by working directly for an agency or by receiving funding from the government – and thus had experience with government regulation over dual use issues. And regardless of whether they are coming from a place of experience or not, scientists worry that government involvement creates more problems than it solves – added regulation slows down science and suppresses innovation. It also creates a hostile environment for scientists to do research. They reflect upon existing mechanisms, like the paperwork and background checks, which are seen as time-consuming and burdensome.

> ***It's very rare that someone would do a malevolent type of research. But I don't think background checks are really a way of finding that out. Because if someone's an alcoholic or if they owe a lot of money, sure, they might do it, but you'd know that anyway. And I think some of the checks we have now are really beyond what is necessary. I think it's almost overblown. [...] It's just something to make people feel a little better.***
> A Life Scientist in San Francisco, California

> ***You'd lose some investigators because they wouldn't want to deal with [more oversight and tighter controls on information dissemination].***
>
> ***You would kill a little bit of serendipity because people wouldn't be able to take advances from one field and apply it to theirs.***
>
> ***Plus, you'd see the best people going elsewhere because there were opportunities, like the stem cell [researchers] going to Europe because they have a different environment for stem cell research.***
>
> ***I think it would inhibit innovation and progress. Maybe not immediately, but in the long term.***
> Life Scientists in Bethesda, Maryland

Regardless, there is tacit agreement inside and outside the Beltway that the government has a role in creating guidelines. However, the sophistication of this understanding varies regionally; in San Francisco there was little knowledge beyond the "NIH", whereas in Bethesda respondents were more aware of other organizations who have a role, like the "DOD" and "NSF".

Participants note government protocols for dual use research in biology are problematic; not all agencies have policies for dual-use research, and those that do often adapt them from other disciplines, which do not necessarily apply in the life sciences. Dual use research sits outside the scope of human subjects and animal use research, leaving program officers to ask the questions informally, and on their own. Unlike the latter two, mechanisms are not in pace to investigate the dual use qualities of new research projects. Participants deeply familiar with the protocols also note that it has been challenging to address the needs of dual-use biological research to policy makers because the existing models for oversight have been based in nuclear research, which uses a different set of assumptions.

We have no official role whatsoever as a program officer. I happen to look carefully at what people are doing, and if I see something that concerns me, I bring it up with them... I'll send whoever it is an e-mail and say I noticed dah, dah, dah, what are you doing about it? Or are you, more often than not, they simply haven't spelled out how they're going to deal with the biosafety issues or haven't clearly told me they've got appropriate facilities. And so I write them and I say tell me how are you going to, where's the work going to be done? Has your IBC approved this work? Is everybody, are you in compliance with select agent regulations and so forth. But we aren't mandated to do that. We are mandated to keep track of human subjects and animal use, whether we're going to go down that road with this kind of stuff, I don't know. It's one possibility. I can't even say really whether I support it or not.
A Life Scientist in Bethesda, Maryland

We don't directly ask the question, is this potential dual use. But we do have in the review process are questions related to does this fit to the mission of the chem/bio defense program?
A Life Scientist in Bethesda, Maryland

I think one of the things that's happened at the DOD is that the people that have been tasked with select agent and the controls come from the nuclear world where they apply those constraints to biologic material. Now, nuclear material decays. It doesn't replicate. So you can't number every single bacteria that you have in your fridge and then trace it back and it's just (inaudible) but the DOD responds to its research priorities.

We were forced, we, as biologists, were forced to use the only system that was in place which was a nuclear system where they could account for certain volumes of radioactivity. When you have a replicating organism, a living replicating organism, it's a little difficult to account for every virus or every bacteria.

Yeah, the whole basis of nuclear control is controlling access to the material and so, you know, for bio, it can't be that. It just can't since you can get the material somewhere else.

Yeah, and a lot of politicians couldn't understand that. They tried to force it on us, force it, no understanding of where the problems lie.
Life Scientists in Bethesda, Maryland

Potential Solutions

Beyond due personal diligence and self-regulation, scientists feel that, beyond relying on the current decentralized system, there are a couple ways scientists themselves can help to increase awareness surrounding dual use issues. First, they feel the science community itself has responsibility, particularly the Institutional review committees (IRCs) that are tasked with overseeing the projects (even though in all groups they neglect to mention that such IRCs may

have people who are not scientists serving on them). The institutional review mechanisms are in place, but they can be improved through increased centralization. They can also take a more active role over dual use research issues. Second, awareness in general should be increased by incorporating into graduate school curricula and training sessions for other scientists greater sensitivity surrounding dual use issues.

Scientists tend to feel that one workable solution would be oversight from other scientists, such as IRCs. However, some point out that IRCs are inconsistent in their practice examples , and that there is a need that something needs to be done to make consistent the IRCs with regards to dual-use protocols.

> ***All pathogen research is reviewed by intramural panel. This panel makes suggestions on the possibility of alternative or safer research, safety measures and guidelines for the publication or omission of sensitive information (such as protocol details).***
> A Life Scientist in Bethesda, Maryland

> ***There is a void right now in if one decided there should be review of things, there is no structure currently in place that's quite the right, I mean, the IBC's are charged with recombinant DNA work. Now most, many universities and probably other institutions have told their IBC's to also deal with biohazard issues. But it's really not, that's not what they've been charged to do. There's no clear, to my knowledge, federal mandate that anything like this be done.***
> A Life Scientist in Bethesda, Maryland

There is a need for widespread education about dual-use issues in the life science community. Though life scientists in these focus groups have had exposure to dual use issues, this awareness comes from the nature of research they conduct or their area of expertise. Most life scientists do not receive an introduction to this in any serious way in graduate school, and some see a role for training programs that can raise awareness.

> ***Pervasive awareness of the dual use issues throughout the life sciences will be critical to minimizing vulnerability. Thus, training programs should be conscientiously developed to help scientist define this term, realize its application, and individually work to minimize their vulnerability.***
> A Life Scientist in Bethesda, Maryland

Appendix A: Methodology

Greenberg Quinlan Rosner Research, Inc. designed and conducted three focus groups of life scientists. Two groups were conducted on February 7, 2007 in Bethesda, MD, and one group was conducted on February 26, 2007 in San Francisco, CA. Participants were recruited by the National Academy of Sciences. Each focus group had a mix of life scientists from different sectors: group 1 in Bethesda was a mixture of scientists from government and academia; group 2 in Bethesda was a mixture of scientists primarily from government and industry; group 3 in San Francisco was a mix of scientists primarily from government and academia. Groups were recruited to have a mix of gender and experience levels. Focus groups are qualitative research and, while indicative, the results are not generalizable.

Appendix B: Author Biographies

Anna Greenberg

Greenberg has extensive experience polling for non-profits and charitable foundations focusing on religion, women's health, rural issues and education.

Prior to joining Greenberg Quinlan Rosner, Greenberg taught at Harvard University's John F. Kennedy School of Government. In the spring of 2000, Greenberg received an invitation from the Pew Research Center for the People and the Press where she worked as a visiting scholar. She serves on the advisory board of the Boisi Center for Religion and American Public Life at Boston College and is a research fellow at American University's Center for Congressional and Presidential Studies.

A frequently quoted source on the topic of American politics, Greenberg has appeared on MSNBC, CNN, NBC, CNBC, NPR and the BBC. Her work has been published in a variety of publications, including Political Science Quarterly, The Chronicle of Higher Education, Women and Politics, The American Prospect, The Nation, Blueprint, The Public Perspective and The Responsive Community.

Greenberg earned a Bachelor's degree from Cornell University and a Ph.D. in Political Science from the University of Chicago.

Contributing Authors

Jennifer Berktold
Jessica Keating

Appendix C

Final Questionnaire

Welcome!

Thank you for your participation in a survey that examines scientists' attitudes toward potential security risks from agricultural, public health, and biomedical research and the role that scientists, institutions, scientific societies, and the government should play in fostering an environment that enhances both the scientific enterprise and national security.

In 2004, the U.S. government established the National Science Advisory Board on Biosecurity (NSABB) under the auspices of the National Institutes of Health to contemplate the possibility and impact of greater oversight for life sciences research to prevent or mitigate deliberate misuse. The NSABB identified several categories of life science experiments that it feels should bear greater scrutiny.

Federal agencies are currently planning to issue further guidelines and considering additional policies regarding responsible scientific research. We believe balancing advancement of the scientific enterprise with the nation's security needs is a very relevant topic to all life scientists. Giving scientists a voice in the policy-making process is one goal of this survey.

The survey is anonymous and will take approximately 20 minutes to complete. Survey results will be shared with policy makers, the scientific community, and the public.

Thank you for your vital contribution to this important policy process.

In recent years, members of the scientific and security policy communities have raised concerns about the potential for misuse of knowledge, tools, and techniques for purposes of bioterrorism. Such research is sometimes called "dual use" research because, although the research is intended for beneficial purposes only, it could be misapplied.

1. Have you ever conducted research or managed others' research in the life sciences?

- ☐ Yes
- ☐ No (goes to question 3)

2. Have you made any changes in how you conduct or manage research because of concerns that knowledge, tools, or techniques from your research might be deliberately misused to facilitate bioterrorism?

	Yes	No
I decided against conducting a specific research project/experiment	☐	☐
I decided to shift my research away from an area altogether	☐	☐
I decided against seeking funding for a proposed research project	☐	☐
I decided against collaborating with particular scientists, postdocs, students, etc.	☐	☐
I limited my conversations about my research	☐	☐
I decided against submitting a manuscript to a journal	☐	☐
I modified a manuscript	☐	☐
I decided against presenting research at a conference	☐	☐
I modified a conference presentation	☐	☐

3. Are you currently conducting or managing research in the life sciences?

- ☐ Yes
- ☐ No (goes to question 5)

In recent years, members of the scientific and security policy communities have raised concerns about the potential for misuse of knowledge, tools, and techniques for purposes of bioterrorism. Such research is sometimes called "dual use" research because, although the research is intended for beneficial purposes only, it could be misapplied.

4. Do you consider any of the research you currently conduct or manage to have dual use potential?

- ☐ Yes
- ☐ No

The National Science Advisory Board for Biosecurity (NSABB) has identified a subset of life sciences research that they believe may be worthwhile but may also need special review. Such research includes experiments designed to (1) enhance the harmful consequences of a biological agent or toxin; (2) disrupt immunity or the effectiveness of an immunization without clinical and/or agricultural justification; (3) confer to a biological agent or toxin, resistance to clinically and/or agriculturally useful prophylactic or therapeutic interventions against that agent or toxin, or facilitate their ability to evade detection methodologies; (4) increase the stability, transmissibility, or the ability to disseminate a biological agent or toxin; (5) alter the host range or tropism of a biological agent or toxin; (6) enhance the susceptibility of a host population; and (7) generate a novel pathogenic agent or toxin, or reconstitute an eradicated or extinct biological agent.

5. Are you currently conducting or managing research that includes any of these seven types of experiments?

- ☐ Yes
- ☐ No

6. Do you now or have you ever worked with or managed research using *select agents*?

- ☐ Yes
- ☐ No
- ☐ Don't know

7. Do the journal(s) in your field require reviewers to evaluate whether manuscripts include knowledge, tools, and techniques with dual use potential?

- ☐ All of the journals have a policy
- ☐ Some of the journals have a policy
- ☐ None of the journals have a policy
- ☐ Don't know

8. Do the journal(s) in your field require authors to disclose any research with dual use potential to editors upon submission of the manuscript?

- ☐ All of the journals have a policy
- ☐ Some of the journals have a policy
- ☐ None of the journals have a policy
- ☐ Don't know

9. Should scientific journals have policies regarding publication of dual use research?

- ☐ Yes
- ☐ No
- ☐ Don't know

10. Have you ever contacted an editor because you felt that a manuscript you were reviewing contained knowledge, tools, or techniques that could pose a threat to national security?

- ☐ Yes
- ☐ No, although I have reviewed manuscripts
- ☐ No, because I have not reviewed manuscripts

11. Should professional science societies have codes for the responsible conduct of dual use life sciences research?

- ☐ Yes
- ☐ No
- ☐ Don't know

12. Are you a member of any professional science societies that have codes of responsible conduct for dual use research?

- ☐ Yes *(please specify)*____
- ☐ No
- ☐ Don't know

13. Principal investigators should be responsible for the initial evaluation of the dual use potential of their life sciences research.

Strongly agree	Agree	Neutral/ No opinion	Disagree	Strongly disagree
☐	☐	☐	☐	☐

14. Scientists should provide formal assurance to their institution that they are assessing their work for dual use potential.

Strongly agree	Agree	Neutral/ No opinion	Disagree	Strongly disagree
☐	☐	☐	☐	☐

15. Scientists conducting or managing research should take an oath, similar to medicine's Hippocratic oath, to carry out research responsibly and guard against deliberate misuse of the knowledge, tools, or techniques of dual use research.

Strongly agree	Agree	Neutral/ No opinion	Disagree	Strongly disagree
☐	☐	☐	☐	☐

16. Preventing the potential that knowledge, tools, or techniques from dual use research could pose a threat to national security requires . . .	SA	A	N/NO	D	SD
Certification of researchers conducting dual use research	☐	☐	☐	☐	☐
Greater restrictions on access to specific biological agents or toxins	☐	☐	☐	☐	☐
Licensure of certain biological equipment that is commonly used in life science research	☐	☐	☐	☐	☐
Restrictions on disclosure of details about the research or its findings through personal communication	☐	☐	☐	☐	☐
Alteration or removal of certain experimental methods or findings prior to publication or presentation	☐	☐	☐	☐	☐
Restrictions on publication of findings based on dual use potential	☐	☐	☐	☐	☐
Classification of research findings based on dual use potential	☐	☐	☐	☐	☐

17. Dual use research needs greater federal oversight.

Strongly agree	Agree	Neutral/ No opinion	Disagree	Strongly disagree
☐	☐	☐	☐	☐

18. Principal investigators should be responsible for training lab staff, students, and visiting scientists about dual use research including policies and practices to minimize the potential for misuse of information from their research.

Strongly agree	Agree	Neutral/ No opinion	Disagree	Strongly disagree
☐	☐	☐	☐	☐

19. University and college students should receive educational lectures and materials on dual use life sciences research including the potential that knowledge, tools, and techniques of such research that could pose a threat to national security.

Strongly agree	Agree	Neutral/ No opinion	Disagree	Strongly disagree
☐	☐	☐	☐	☐

20. Institutions should provide mandatory training for scientists regarding dual use life sciences research.

Strongly agree	Agree	Neutral/ No opinion	Disagree	Strongly disagree
☐	☐	☐	☐	☐

21. All grant proposals for life sciences research with dual use potential should be reviewed by a researcher's institution prior to submission for funding.

Strongly agree	Agree	Neutral/ No opinion	Disagree	Strongly disagree
☐	☐	☐	☐	☐

22. Funding agencies should require grantees to attest on grant applications that they have considered dual use implications of their proposed research.

Strongly agree	Agree	Neutral/ No opinion	Disagree	Strongly disagree
□	□	□	□	□

23. Funding agencies would be less likely to fund grant proposals if the proposed research had dual use potential.

Strongly agree	Agree	Neutral/ No opinion	Disagree	Strongly disagree
□	□	□	□	□

24. What is the percent chance (ranging from 0 percent chance to 100 percent chance) that an act of bioterrorism will occur somewhere in the world in the next 5 years?

_________ %

25. What is the percent chance (ranging from 0 percent chance to 100 percent chance) that an act of bioterrorism will occur in the United States in the next 5 years?

_________ %

26. What is the percent chance (ranging from 0 percent chance to 100 percent chance) that knowledge, tools, or techniques from dual use life sciences research will facilitate an act of bioterrorism in the next 5 years?

_________ %

27. To date, there have been few acts of bioterrorism. Which of the following help explain why?

	Yes	No	Don't Know
Terrorists lack the knowledge to work with or create dangerous biological agents.	□	□	□
Terrorists lack the equipment to work with or create dangerous biological agents.	□	□	□

Terrorists lack access to dangerous biological agents.	☐	☐	☐
Terrorists are deterred by the threat of being caught and punished.	☐	☐	☐
Terrorists prefer to use other means.	☐	☐	☐

28. Do the following means of communication provide sufficient information for an individual with college-level life science training to deliberately create a harmful biological agent?

	Yes	No	Don't Know
Scientific journal articles	☐	☐	☐
Presentations at scientific conferences or meetings	☐	☐	☐
Personal communications (e.g., e-mail, phone calls)	☐	☐	☐
Internet	☐	☐	☐

29. On June 1, 2007, what was your citizenship status?

- ☐ U.S. citizen, since birth or naturalized
- ☐ non-U.S. citizen, with a Permanent U.S. Resident Visa (Green Card)
- ☐ non-U.S. citizen, with a Temporary U.S. Resident Visa

30. What is the highest educational degree you have been awarded?

- ☐ Bachelor's degree or equivalent (e.g., B.S., B.A., A.B.)
- ☐ Master's degree or equivalent (e.g., M.S., M.A., M.B.A.)
- ☐ Doctorate or equivalent (e.g., Ph.D., D.Sc., Ed.D., etc.)
- ☐ Other professional degree (e.g., J.D., L.L.B., M.D., D.D.S., D.V.M., etc.)
- ☐ Joint doctorate and professional degree (e.g., Ph.D. and M.D.)
- ☐ Other

31. In what year was your highest educational degree awarded?

(YYYY)

32. Which one of the following best describes your current occupational status? Are you . . .

- ☐ Employed
- ☐ Unemployed [goes to question 34]
- ☐ Retired [goes to question 34]
- ☐ Other—please specify [goes to question 34] ______________

33. Which one of the following best describes your principal employer during the week of June 1, 2007?

- ☐ Industry (including self-employed, business owner, private sector employee)
- ☐ Educational institution
- ☐ Government employee (including city, county, federal, or military service)
- ☐ Other

34. Which scientific discipline of the following do you consider to be your primary area of work or study? (If currently unemployed or retired, please select the discipline that most closely matches your last occupation.)

- ☐ Agricultural Science
- ☐ Biochemistry
- ☐ Biomedical Engineering
- ☐ Biotechnology
- ☐ Botany
- ☐ Cell Biology
- ☐ Ecology
- ☐ Endocrinology/Physiology
- ☐ Genetics
- ☐ Geology/Soil Sciences/Geography
- ☐ Immunology
- ☐ Marine Biology
- ☐ Medicine
- ☐ Microbiology
- ☐ Molecular Biology
- ☐ Neuroscience
- ☐ Pharmacology
- ☐ Zoology
- ☐ Other: ______________________

35. Do you have any additional comments regarding regulation and oversight of dual use research you would like federal policy makers to consider? (optional)

Thank you very much for taking the time to complete this survey!

Appendix D

Additional Data and Analysis

TABLE D-1 Survey Questions and Number of Responses for Each Question

Question	Number of Responses[a]
1. Have you ever conducted research or managed others' research in the life sciences?	1,950
2. Have you made any changes in how you conduct or manage research because of concerns that knowledge, tools, or techniques from your research might be deliberately misused to facilitate bioterrorism?	
I decided against conducting a specific research project/experiment	1,744
I decided to shift my research away from an area altogether	1,744
I decided against seeking funding for a proposed research project	1,744
I decided against collaborating with particular scientists, postdocs, students, etc.	1,744
I limited my conversations about my research	1,744
I decided against submitting a manuscript to a journal	1,744
I modified a manuscript	1,744
I decided against presenting research at a conference	1,744
I modified a conference presentation	1,744
3. Are you currently conducting or managing research in the life sciences?	1,843
4. Do you consider any of the research you currently conduct or manage to have dual use potential?	1,376[b]
5. Are you currently conducting or managing research which includes any of these seven types of experiments?	1,376[c]

Question	Number of Responses[a]
6. Do you now or have you ever worked with or managed research using select agents?	1,798
7. Do the journal(s) in your field require reviewers to evaluate whether manuscripts include knowledge, tools and techniques with dual use potential?	1,755
8. Do the journal(s) in your field require authors to disclose any research with dual use potential to editors upon submission of the manuscript?	1,755
9. Should scientific journals have policies regarding publication of dual use research?	1,755
10. Have you ever contacted an editor because you felt that a manuscript you were reviewing contained knowledge, tools, or techniques that could pose a threat to national security?	1,755
11. Should professional scientific societies have codes for the responsible conduct of dual use life sciences research?	1,743
12. Are you a member of any professional scientific societies that already have codes of conduct that include statements about the responsible conduct of dual use research?	1,743
13. Principal investigators should be responsible for the initial evaluation of the dual use potential of their life sciences research.	1,658
14. Scientists should provide formal assurance to their institution that they are assessing their work for dual use potential.	1,658
15. Scientists conducting or managing research should take an oath, similar to medicine's Hippocratic Oath, to carry out research responsibly and guard against deliberate misuse of the knowledge, tools, or techniques of dual use research.	1,658
16. Preventing the potential that knowledge, tools, or techniques from dual use research could pose a threat to national security requires . . .	1,658
Certification of researchers conducting dual use research	1,658
Greater restrictions on access to specific biological agents or toxins.	1,658
Licensure of certain biological equipment that is commonly used in life science research.	1,658
Restrictions on disclosure of details about the research or its findings through personal communication.	1,658
Alteration or removal of certain experimental methods or findings prior to publication or presentation.	1,658
Restrictions on publication of findings based on dual use potential.	1,658
Classification of research findings based on dual use potential.	1,658
17. Dual use research needs greater federal oversight.	1,637
18. Principal investigators should be responsible for training lab staff, students, and visiting scientists about dual use research including policies and practices to minimize the potential for misuse of information from their research.	1,637

Question	Number of Responses[a]
19. University and college students should receive educational lectures and materials on dual use life sciences research including the potential that knowledge, tools, and techniques from such research could pose a threat to national security.	1,637
20. Institutions should provide mandatory training for scientists regarding dual use life sciences research.	1,637
21. All grant proposals for life sciences research with dual use potential should be reviewed by a researcher's institution prior to submission for funding.	1,633
22. Funding agencies should require grantees to attest on grant applications that they have considered dual use implications of their proposed research.	1,633
23. Funding agencies would be less likely to fund grant proposals if the proposed research has dual use potential.	1,633
24. What is the percent chance (ranging from 0 percent chance to 100 percent chance) that an act of bioterrorism will occur somewhere in the world in the next five years?	1,588
25. What is the percent chance (ranging from 0 percent chance to 100 percent chance) that an act of bioterrorism will occur in the United States in the next five years?	1,588
26. What is the percent chance (ranging from 0 percent chance to 100 percent chance) that knowledge, tools, or techniques from dual use life sciences research will facilitate an act of bioterrorism in the next five years?	1,588
27. To date, there have been few acts of bioterrorism. Which of the following help explain why?	
Terrorists lack the knowledge to work with or create dangerous biological agents.	1,588
Terrorists lack the equipment to work with or create dangerous biological agents.	1,588
Terrorists lack access to dangerous biological agents.	1,588
Terrorists are deterred by the threat of being caught and punished.	1,588
Terrorists prefer to use other means.	1,588
28. Do the following means of communication provide sufficient information for an individual with college level life science training to deliberately create a harmful biological agent?	
Scientific journal articles	1,588
Presentations at scientific conferences or meetings	1,588
Personal communications (e.g., e-mail, phone calls)	1,588
Internet	1,588
29. On June 1, 2007, what was your citizenship status?	1,586
30. What is the highest educational degree you have been awarded?	1,586

Question	Number of Responses[a]
32. Which one of the following best describes your current occupational status? Are you . . .	1,586
33. Which one of the following best describes your principal employer during the week of June 1, 2007?	1,443
34. Which scientific discipline of the following do you consider to be your primary area of work or study? (If currently unemployed or retired, please select the discipline that most closely matches your last occupation.)	1,586

[a] Unless otherwise noted, these numbers are the number of responses to each question out of the possible 1,954 respondents who answered at least part of the survey.

[b]This number is the number of respondents who answered this question out of the 1,407 respondents who answered "yes" to question #3.

[c]This number is the number of respondents who answered this question out of the 1,407 respondents who answered "yes" to question #3.

TABLE D-2 Percentage Likelihood of Dual Use Research Facilitating a Bioterror Attack, by Type of Research in Which Respondents Are Engaged

	Percentage Likelihood of Dual Use Facilitating a Bioterror Attack		
Type of Research	Mean (%)	SD (%)	N
Works with dual use			
Yes	28	33	196
No	25	30	1,033
Total			1,229[a]
Works with seven types of experiments			
Yes	31	33	74
No	25	30	1,155
			1,229[a]
Works with select agents			
Yes	30	34	416
No	27	30	1,051
Don't Know	29	32	120
Total			1,587[b]

[a]These two questions (works with dual use and percent likelihood and works with seven types of experiments and percent likelihood) could only be answered by the 1,407 people who were currently engaged in research. The 178 other respondents did not answer one or both of these questions.

[b]The questions about working with select agents and percent likelihood were asked of all 1,954 respondents. The other 367 respondents failed to answer one or both of these questions.

SOURCE: NRC/AAAS Survey; data tabulations by staff.

TABLE D-3 Percentage of Respondents Agreeing or Disagreeing with Statement That Greater Federal Oversight Is Needed, by Type of Research and Employment

Variable	Values	Strongly Disagree	Disagree	Neutral	Agree	Strongly Agree	Observations
Works with dual use	Yes	20	40	16	21	3	198
	No	11	35	29	21	4	1,065
Works with seven types of experiments	Yes	23	32	18	22	5	74
	No	12	36	28	21	4	1,189
Works with select agents	Yes	15	40	21	19	4	427
	No	10	34	30	23	4	1,088
Employer type	Industry	9	37	23	25	7	223
	Academe	12	36	28	20	3	1,023
	Government	10	32	28	26	4	125
	Other	14	29	25	29	3	72

NOTES: The number of observations reflects the number of individuals who answered each question down the left column ("Variable") and the likert scale question. Recall that only 1,407 individuals were asked about whether they considered their research to be dual use or involve the seven types of experiments. As can be seen in the far-right column, not all of these 1,407 individuals answered the combination of dual use research and the likert scale question or seven types of experiments and the likert scale question. All 1,954 respondents could have answered whether they work with select agents and what their employer type was, although as noted in the far-right column, not all 1,954 respondents actually did so.
SOURCE: NRC/AAAS Survey; data tabulations by staff.

TABLE D-4 Percentage of Respondents Agreeing or Disagreeing That Particular Policies Should Be Required, by Policy, Type of Research and Employment

Variable	Values	Strongly Disagree	Disagree	Neutral	Agree	Strongly Agree	Observations
A. Certification of researchers							
Works with dual use	Yes	17	35	9	28	11	198
	No	10	25	24	31	11	1,077
Works with seven types of experiments	Yes	20	30	11	30	9	74
	No	10	26	22	31	11	1,201
Works with select agents	Yes	14	27	17	29	13	429
	No	10	25	23	32	10	1,100
Employer type	Industry	12	26	18	35	9	223
	Academia	11	26	22	30	11	1,023
	Government	9	23	27	29	12	125
	Other	11	24	28	29	8	72

Variable	Values	Strongly Disagree	Disagree	Neutral	Agree	Strongly Agree	Observations
B. Restrictions on access							
Works with dual use	Yes	19	30	9	33	10	198
	No	9	24	22	34	12	1,077
Works with seven types of experiments	Yes	18	32	4	36	9	74
	No	10	24	21	34	11	1,201
Works with select agents	Yes	15	28	16	28	13	429
	No	7	22	21	38	13	1,100
Employer type	Industry	7	21	15	40	17	223
	Academia	10	26	20	34	11	1,023
	Government	7	23	13	40	17	125
	Other	7	15	35	35	8	72
C. Licensure of equipment							
Works with dual use	Yes	32	42	9	14	4	198
	No	23	37	21	14	4	1,077
Works with seven types of experiments	Yes	32	42	9	15	1	74
	No	24	38	20	14	4	1,201
Works with select agents	Yes	26	39	17	13	5	429
	No	22	35	21	17	5	1,100
Employer type	Industry	22	35	18	19	6	223
	Academia	25	39	19	14	4	1,023
	Government	18	38	19	19	6	125
	Other	19	32	25	21	3	72
D. Restrictions on personal communication							
Works with dual use	Yes	33	32	15	18	2	198
	No	20	38	23	16	4	1,077
Works with seven types of experiments	Yes	22	34	18	26	1	74
	No	22	37	22	15	4	1,201
Works with select agents	Yes	27	33	18	17	4	429
	No	19	37	24	17	3	1,100
Employer type	Industry	18	31	24	21	6	223
	Academia	22	39	22	15	3	1,023
	Government	23	26	26	19	5	125
	Other	24	33	24	15	4	72
E. Modification of manuscripts or presentations							
Works with dual use	Yes	33	26	15	23	3	198
	No	22	35	23	17	3	1,077
Works with seven types of experiments	Yes	26	30	12	31	1	74
	No	24	34	22	17	3	1,201
Works with select agents	Yes	25	32	20	19	5	429
	No	21	35	23	19	3	1,100

Variable	Values	Strongly Disagree	Disagree	Neutral	Agree	Strongly Agree	Observations
Employer type	Industry	18	29	22	24	6	223
	Academia	24	36	20	17	3	1,023
	Government	21	30	26	19	4	125
	Other	19	35	25	19	1	72
F. Restrictions on publications							
Works with dual use	Yes	28	36	16	17	3	198
	No	20	38	23	17	3	1,077
Works with seven types of experiments	Yes	28	42	15	12	3	74
	No	20	37	23	17	3	1,201
Works with select agents	Yes	19	36	23	19	3	429
	No	18	37	24	19	2	1,100
Employer type	Industry	16	29	26	25	5	223
	Academia	21	40	21	16	2	1,023
	Government	19	35	25	18	3	125
	Other	19	39	21	21	0	72
G. Classification of findings							
Works with dual use	Yes	32	27	21	18	3	198
	No	19	28	26	24	3	1,077
Works with seven types of experiments	Yes	27	28	27	18	0	74
	No	21	28	25	23	3	1,201
Works with select agents	Yes	24	25	21	26	4	429
	No	17	29	27	24	4	1,100
Employer type	Industry	15	25	24	28	8	223
	Academia	20	31	25	22	3	1,023
	Government	22	19	26	27	6	125
	Other	18	22	28	31	1	72

NOTES: The number of observations reflects the number of individuals who answered each question down the left column ("Variable") and the likert scale question. Recall that only 1,407 individuals were asked about whether they considered their research to be dual use or involve the seven types of experiments. As can be seen in the far-right column, not all of these 1,407 individuals answered the combination of dual use research and the likert scale question or seven types of experiments and the likert scale question. All 1,954 respondents could have answered whether they work with select agents and what their employer type was, although as noted in the far-right column, not all 1,954 respondents actually did so.
SOURCE: NRC/AAAS Survey; data analysis by staff.